this book belong to

_____

_____

1st Grade Common Core Math

*Write the number of dots. Find 1 or 2 groups that make finding the total number of dots easier!

1	●●		16	●●●●● ●●●●	
2	●●●		17	●●●●● ●●●	
3	●●●●		18	●●●●● ●●●●●	
4	●●●		19	●●●●● ●●	
5	●		20	●●●●● ●	
6	●●●●		21	●●●●● ●●●●	
7	●●●●●		22	●●●●● ●●●●●	
8	●●●●		23	●●●● ●●●●●	
9	●●●●● ●		24	●●●●● ●●●	
10	●●●●● ●●		25	●●● ●● ●●●●●	
11	●●●●●		26	●●●●● ●●	
12	●●●●		27	●●● ●● ●● ●●●	
13	●●●●● ●		28	●● ●● ●● ●●	
14	●●●●● ●●●		29	●● ●●● ●● ●	
15	●●●●● ●●		30	●● ●● ●● ●●●	

*Write the number of dots. Find 1 or 2 groups that make finding the total number of dots easier!

1	•		16	••••• •••	
2	••		17	••••• ••••	
3	•		18	••••• ••	
4	••••		19	••••• •••	
5	•••		20	••••• •••••	
6	•••••		21	••••• ••••	
7	••••		22	••••• •••••	
8	•••••		23	• •••• •••••	
9	••••• ••		24	••••• •••••	
10	••••• •		25	•• •••••	
11	••••• •••		26	••• • •• ••	
12	••••• •		27	•• ••• ••• ••	
13	•••••		28		
14	••••• ••		29		
15	••••• •		30		

Circle 5 and make a number bond.

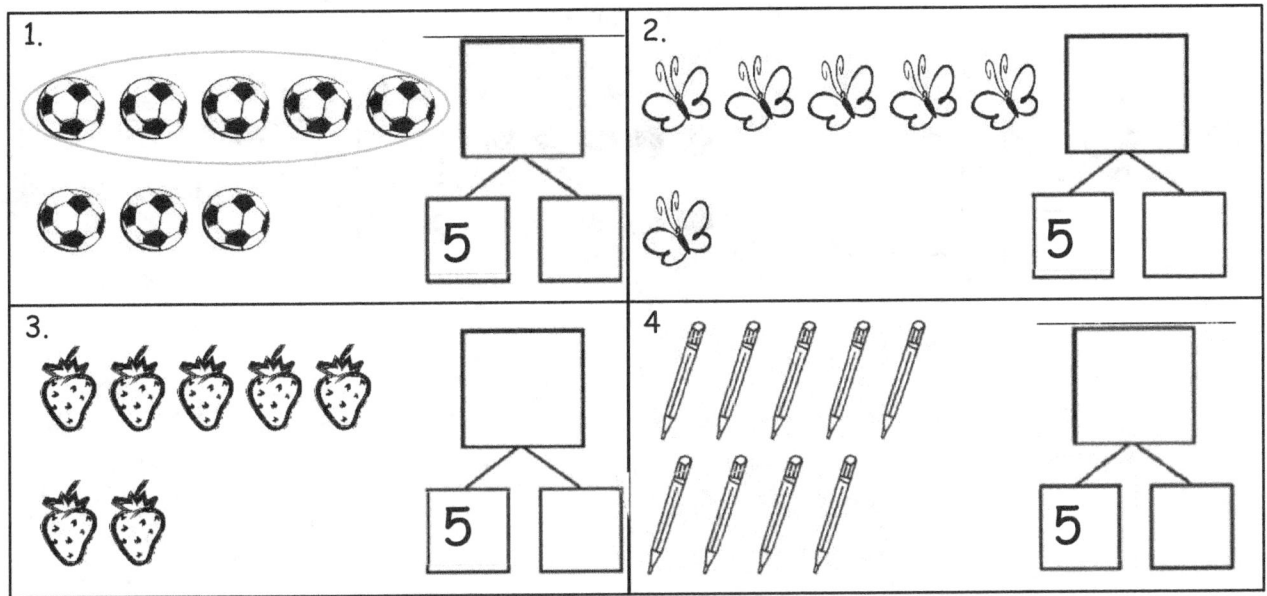

Put nail polish on the number of fingernails shown from left to right. Then fill in the parts. Make the number of fingernails on one hand a part.

5.

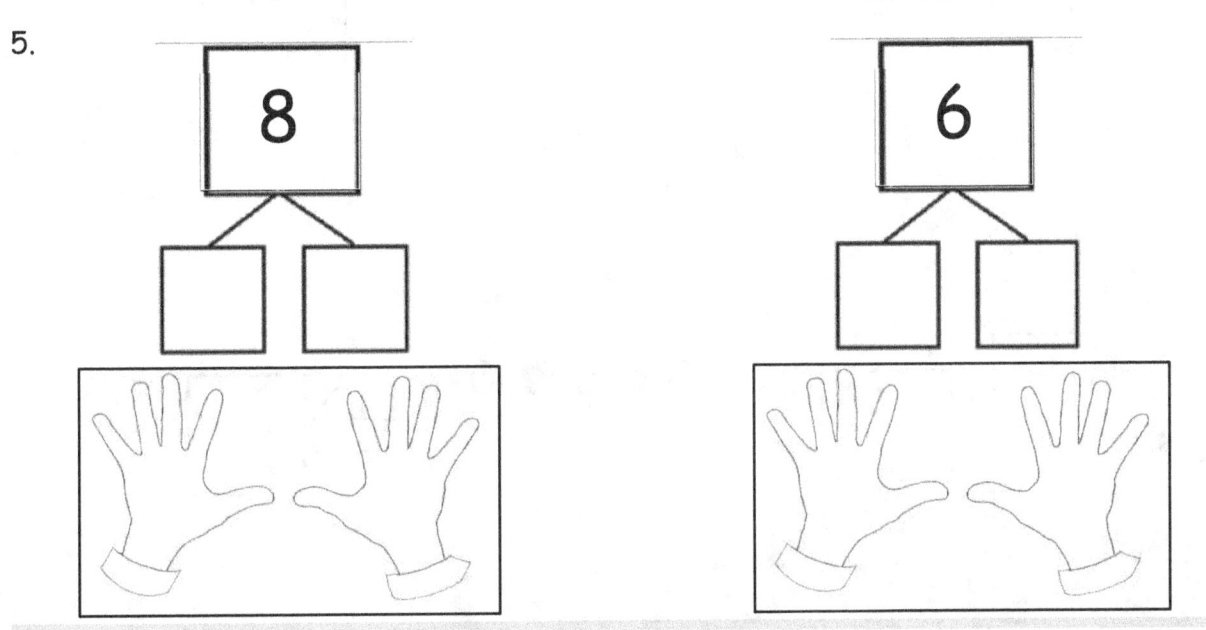

6. Make a number bond that shows 5 as one part.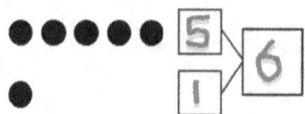

7.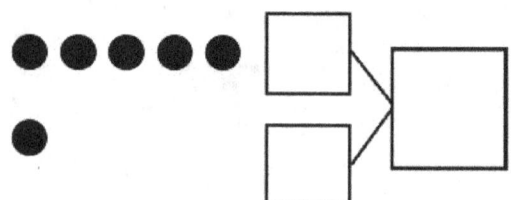

8.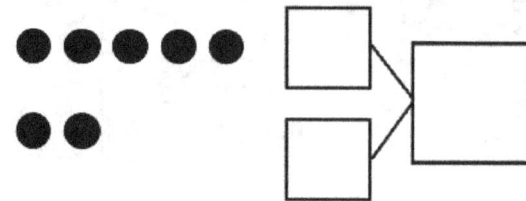

9.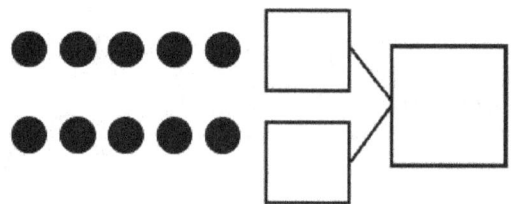

10.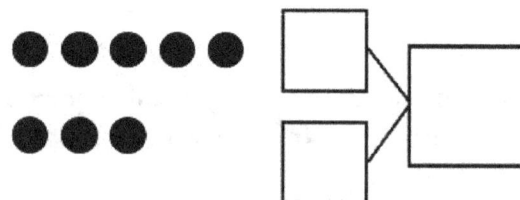

11.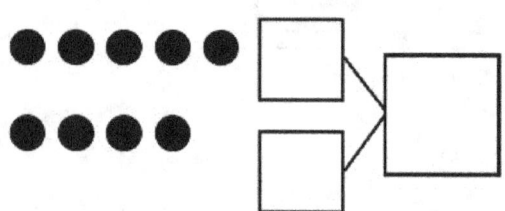

12.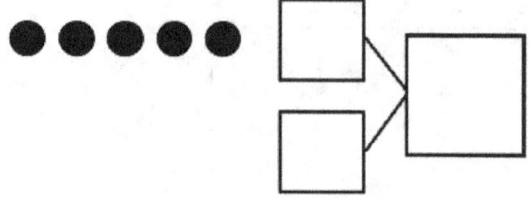

Number Bond Dash!

<u>Directions</u>: Do as many as you can in 60 seconds. Write the amount you finished here:

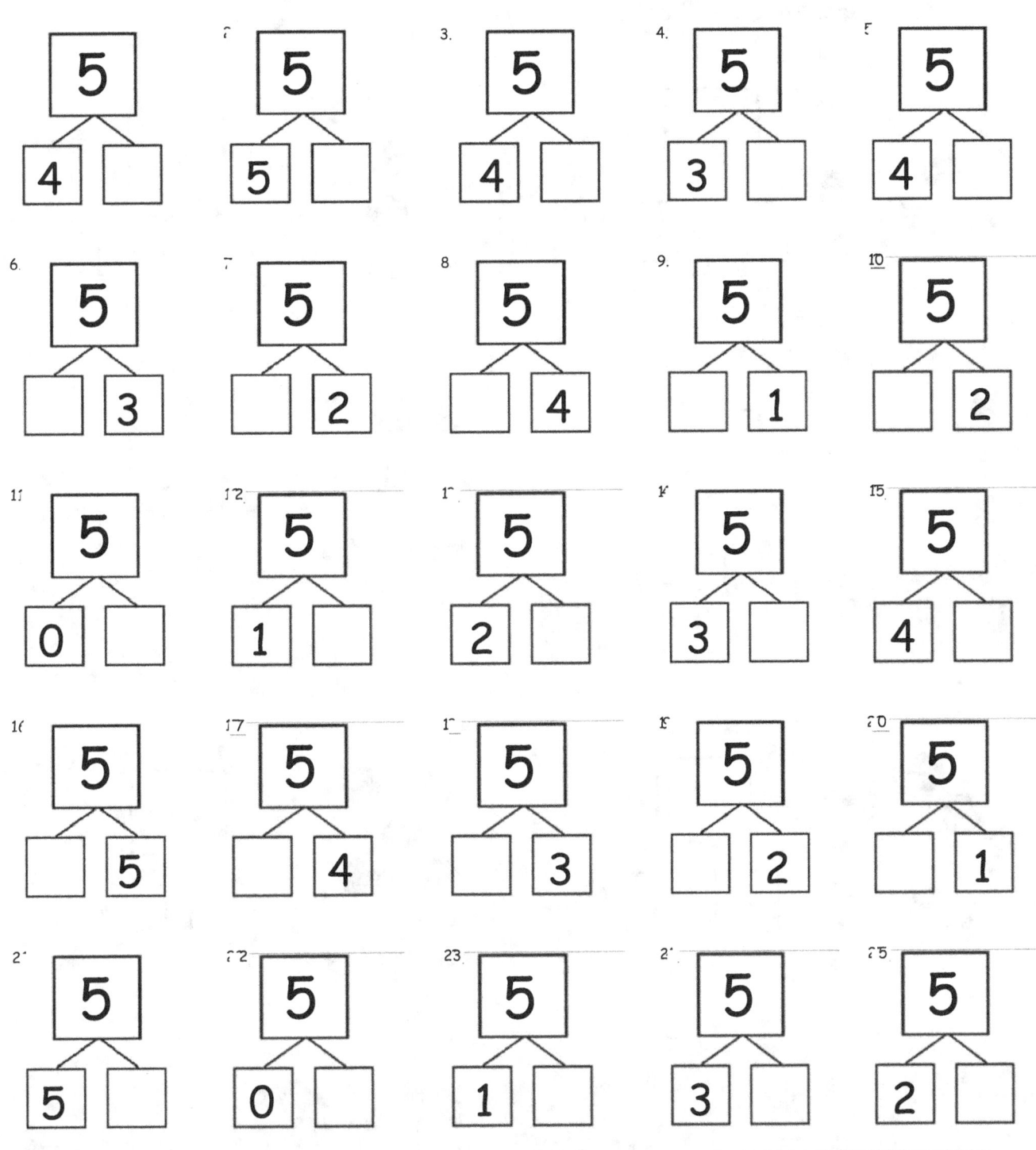

Circle 2 parts you see. Make a number bond to match.

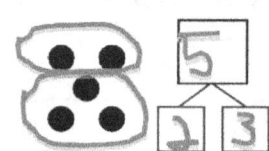

1.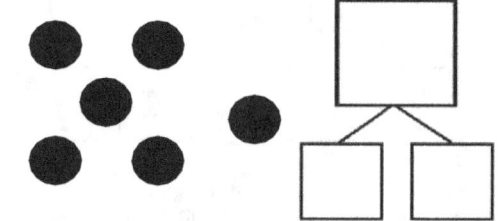

2.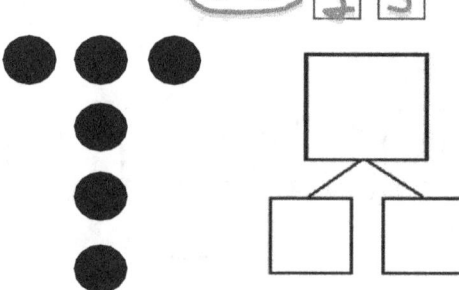

3.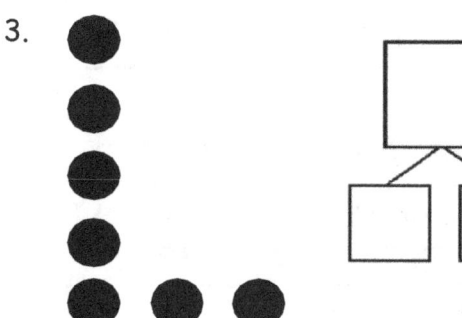

4.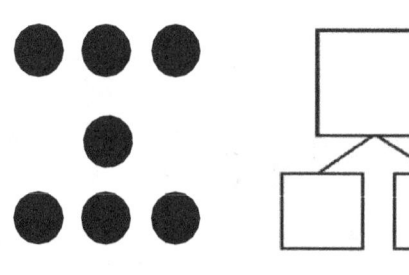

5.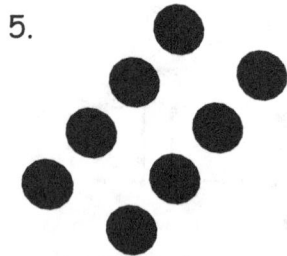

6.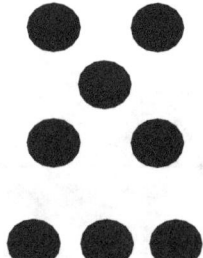

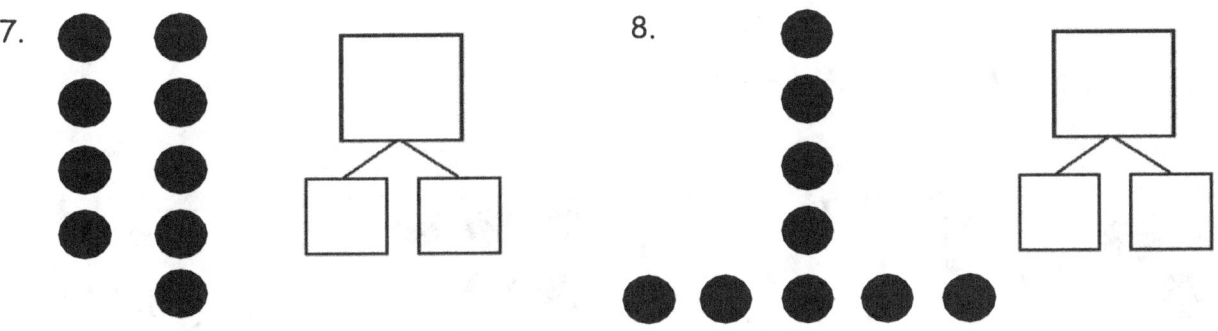

9. How many pieces of fruit do you see? Write at least 2 different number bonds to show different ways to break apart the total.

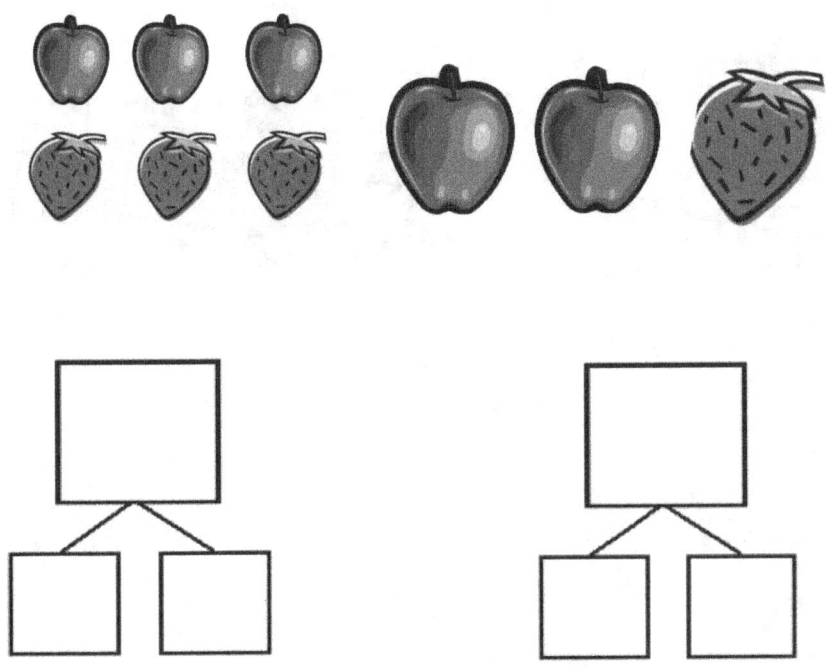

Circle 2 parts you see. Make a number bond to match.

1.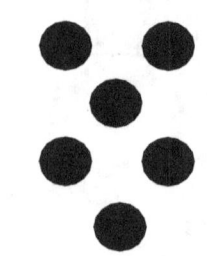

2.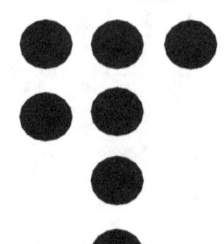

3.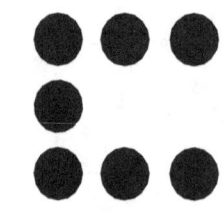

4.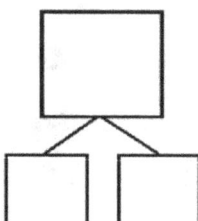

Draw one more in the 5-group. In the box, write the numbers to describe the new picture.

1.

   1 more than 7 is_____.

   7 + 1 = _____

2.

   1 more than 9 is_____.

   9 + 1 = _____

3.

   1 more than 6 is_____.

   6 + 1 = _____

4.

   1 more than 5 is_____.

   5 + 1 = _____

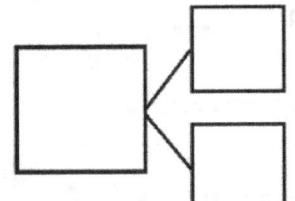

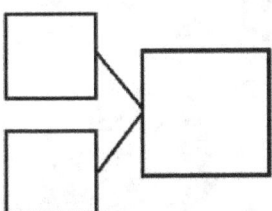

5.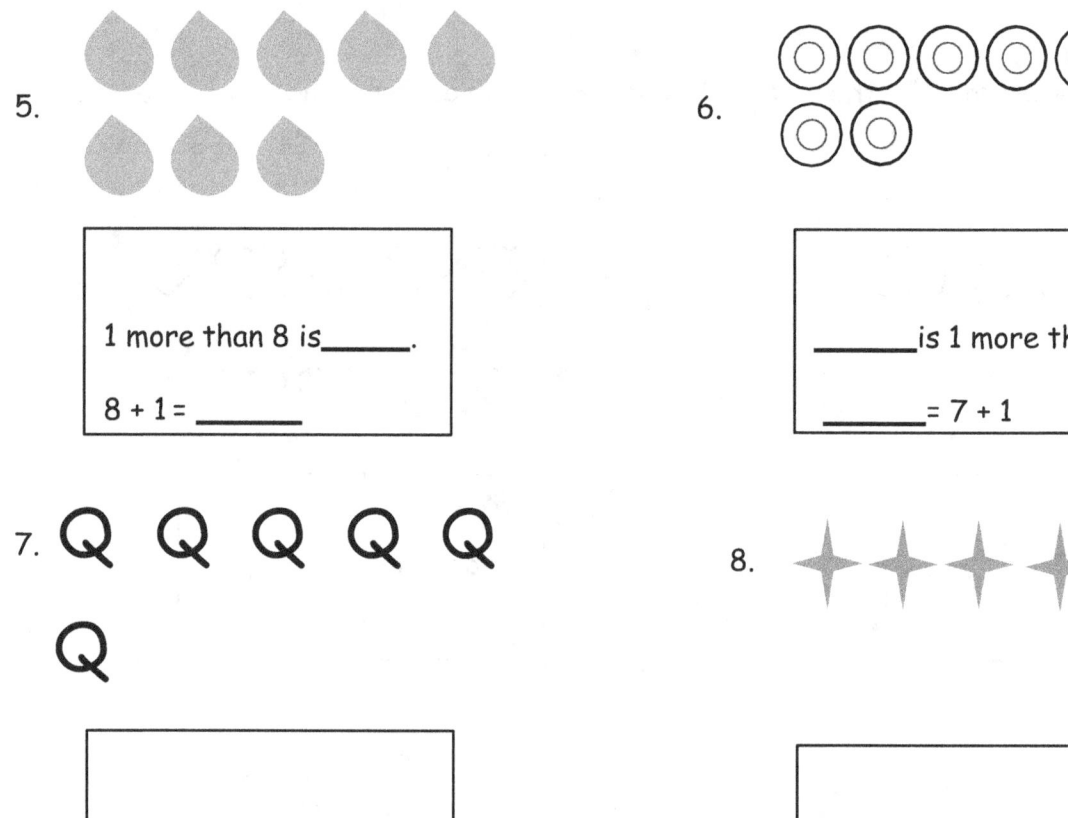

1 more than 8 is _____.

8 + 1 = _____

6.

_____ is 1 more than 7

_____ = 7 + 1

7.

_____ is 1 more than 6

_____ = 6 + 1

8.

_____ is 1 more than 5.

_____ = 5 + 1

9. Imagine adding 1 more backpack to the picture. Then write the numbers to match how many backpacks there will be.

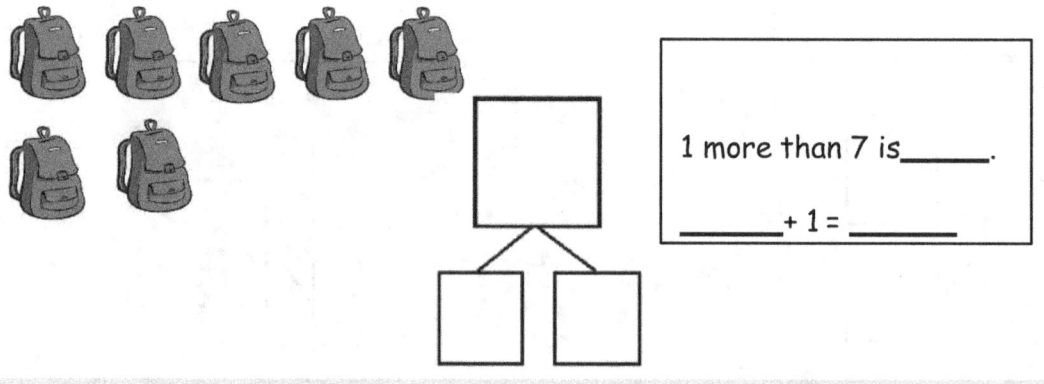

1 more than 7 is _____.

_____ + 1 = _____

Name _____   Date _____

How many objects do you see? Draw one more. How many objects are there now?

1.

_____ is 1 more than 9.

9 + 1 = _____

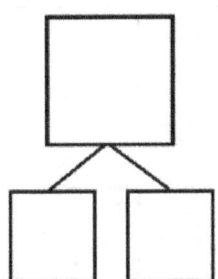

2.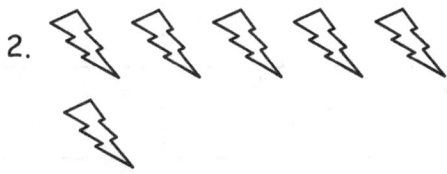

1 more than 6 is _____.

_____ + 1 = _____

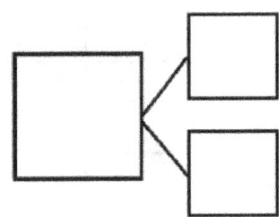

*Write the number that is 1 more.

	•••		••••• ••••	
	••		9	
	•••		7	
	••••		••••• ••	
	•••••		8	
	••••• •		7	
	•••••		••••• •••	
	5		••••• ••••	
	••••• ••		10	
	6		••••• •••••	
	••••• •		••••• •••	
	7		•• •• •• ••	
	••••• ••		9	
	••••• •••		••• ••• •••	
	8		••• ••• ••• •••	

Ways to Make 6!

Use the apple picture to help you write all of the different ways to make 6.

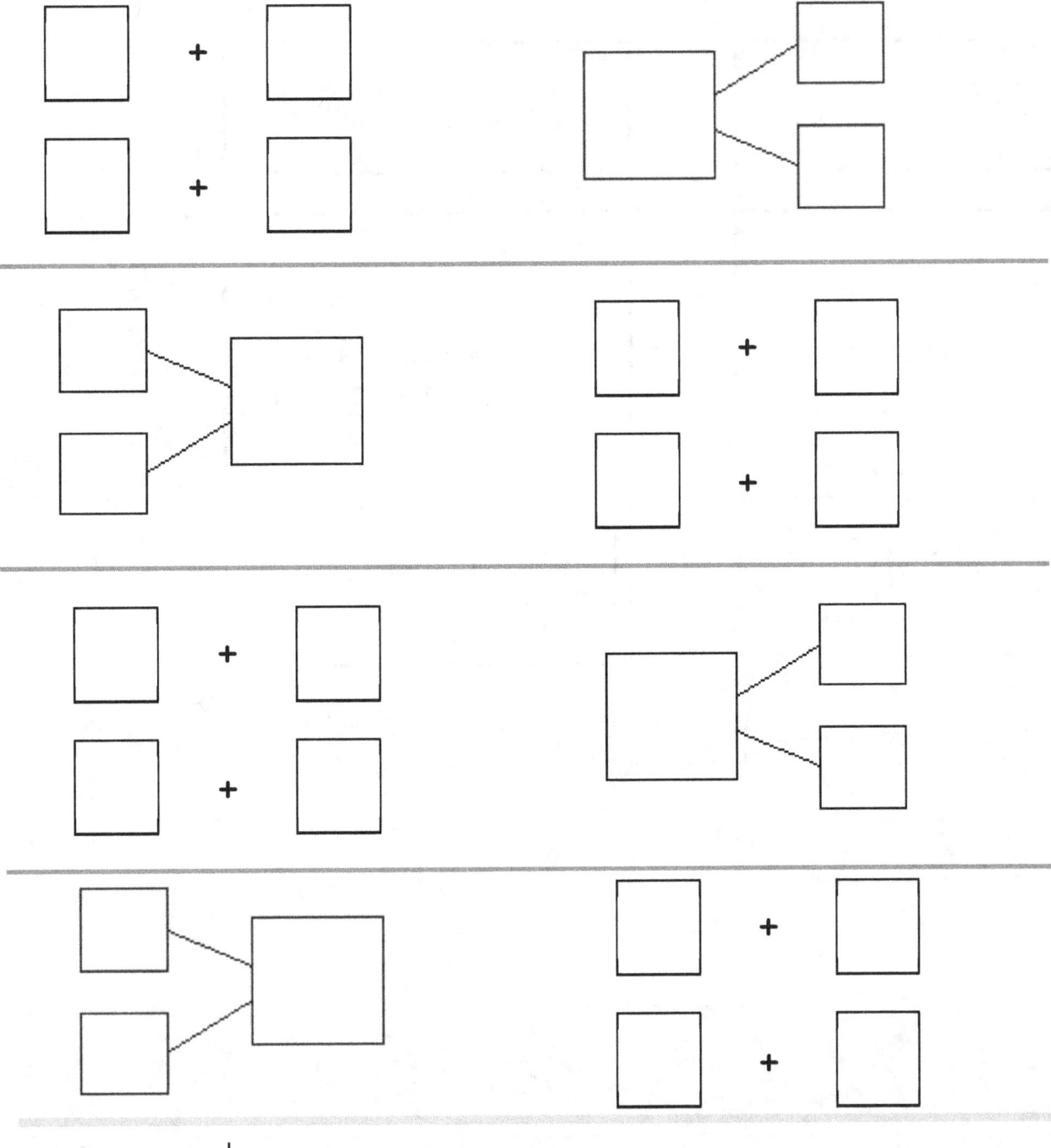

Show different ways to make 6. In each set, shade some circles and leave the others blank.

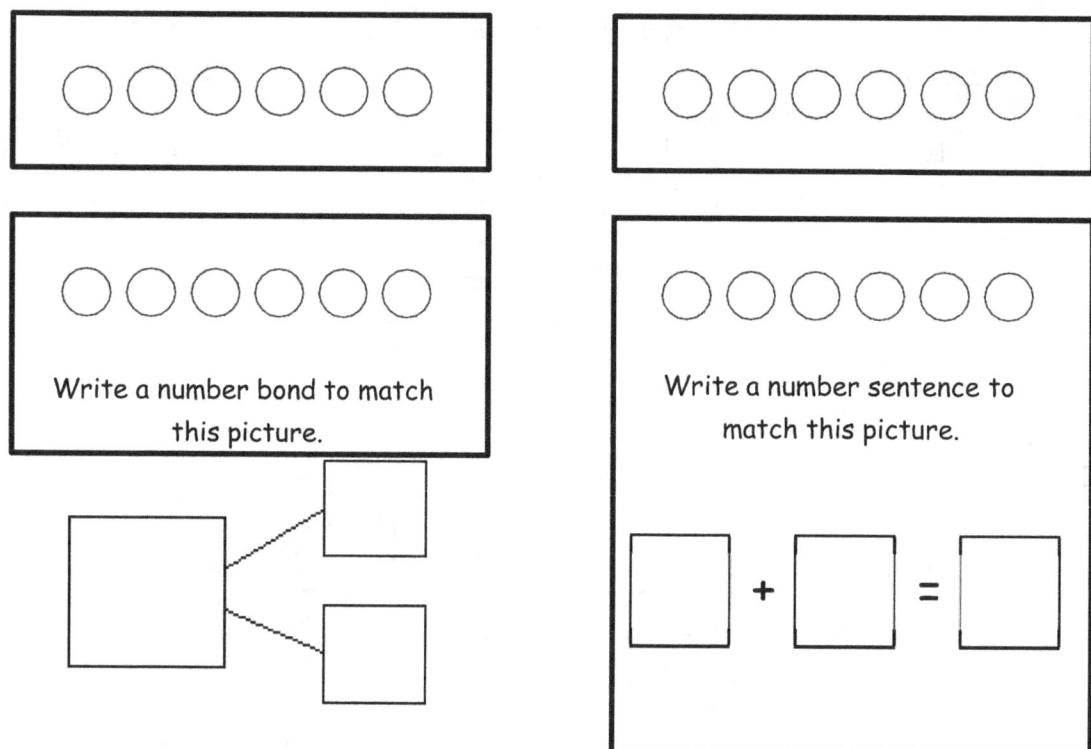

Write a number bond to match this picture.

Write a number sentence to match this picture.

Shake Those Disks! - 6

6	6	6	6
0 6	1 5	2 4	3 3

Number Bond Dash!

<u>Directions</u>: Do as many as you can in 90 seconds. Write the amount you finished here:

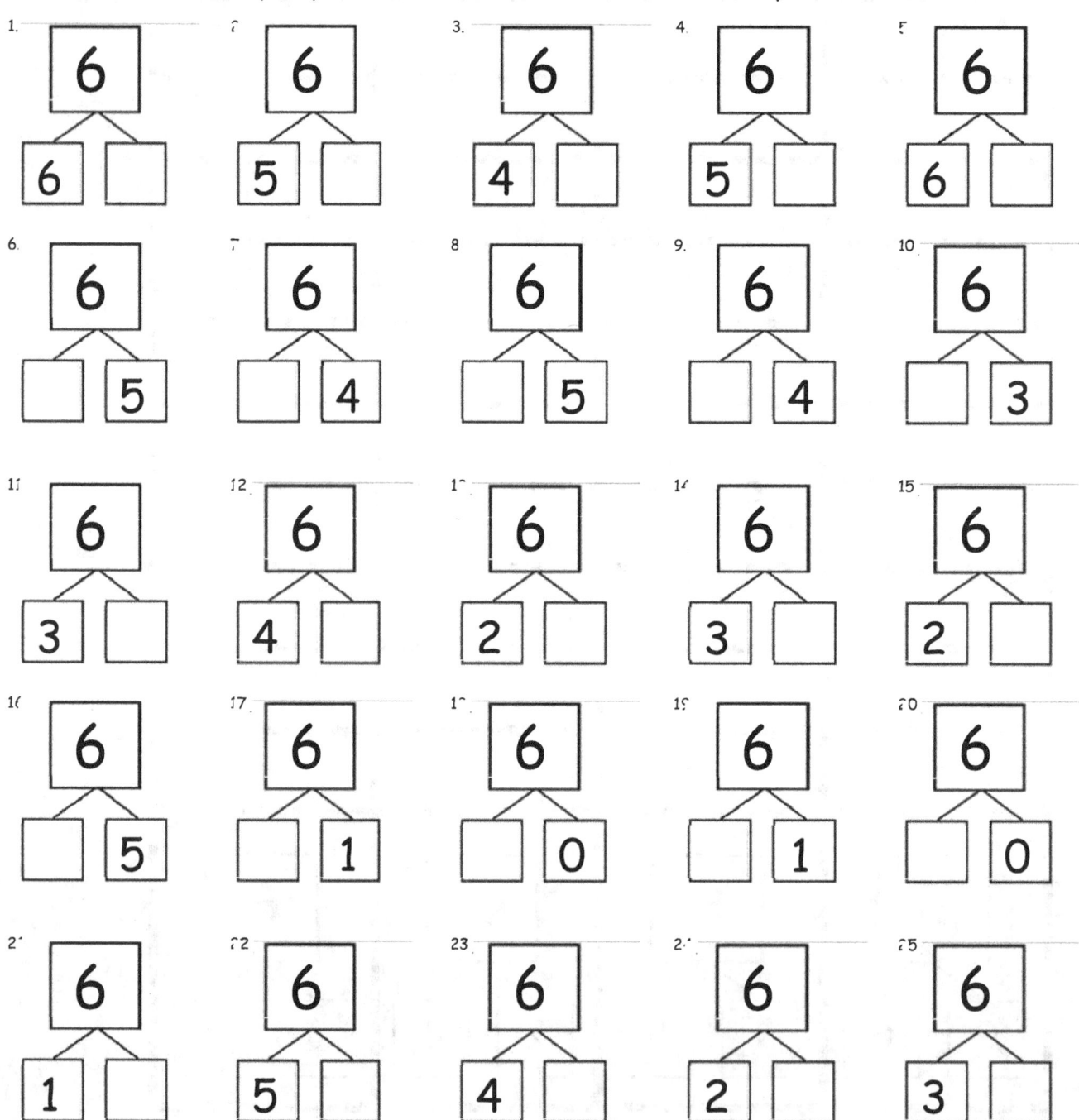

Ways to Make 7! Use the classroom picture to help you write the expressions and number bonds to show all of the different ways to make 7.

Color in two dice that make 7 together. Then fill in the number bond and number sentences to match the dice you colored.

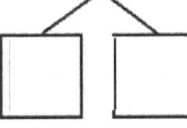

 ○  = 7   7 =

 ○  = 7   7 = 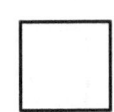 ○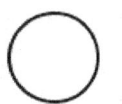

Number Bond Dash!

<u>Directions</u>: Do as many as you can in 90 seconds. Write the amount you finished here: _____

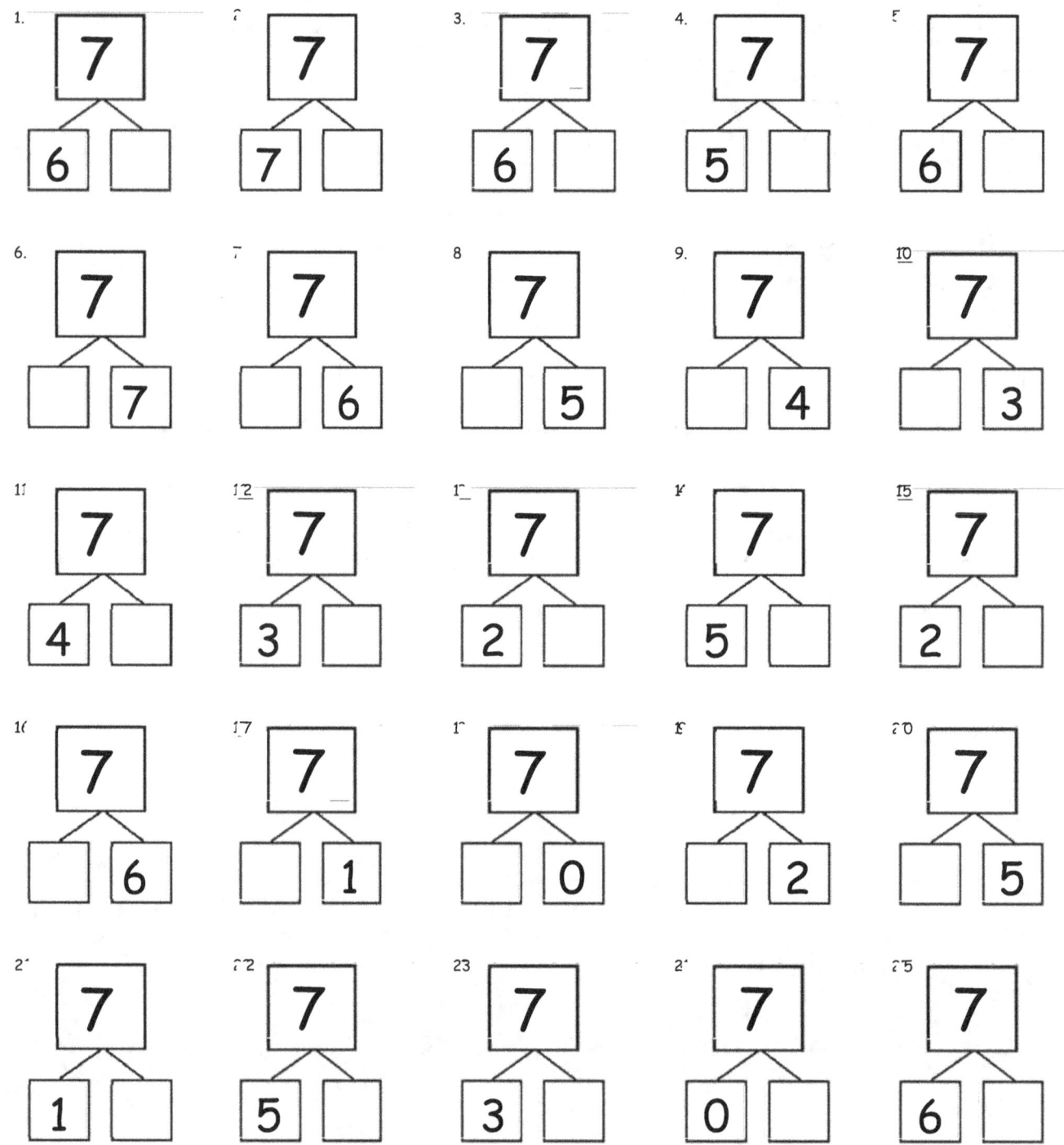

Ways to Make 8 Game Recording Sheet

Use your 5-group cards to help you write the expressions and number bonds to show all of the different ways to make 8.

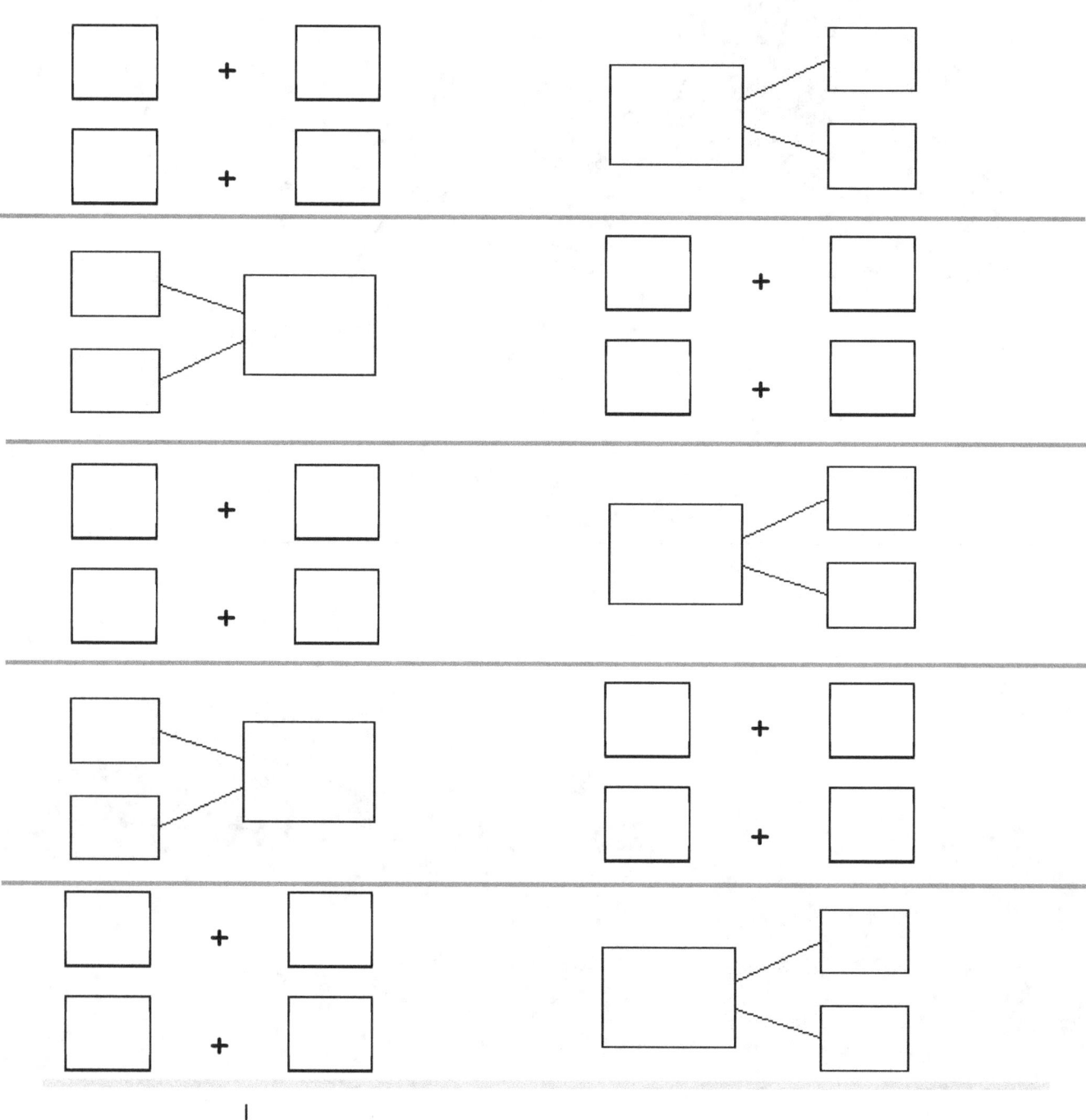

Circle the part. Count on to show 8 with the picture and number bond. Write the expressions.

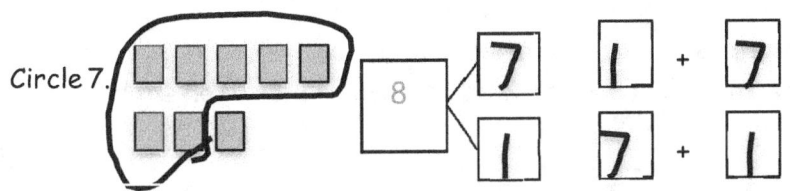

1. Circle 6. How many more does 6 need to make 8?

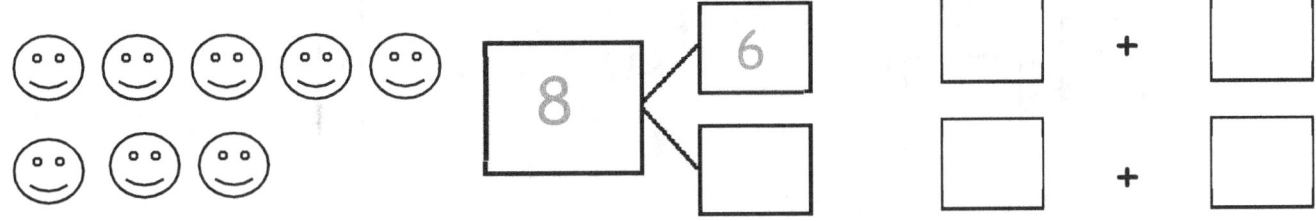

2. Circle 5. How many more does 5 need to make 8?

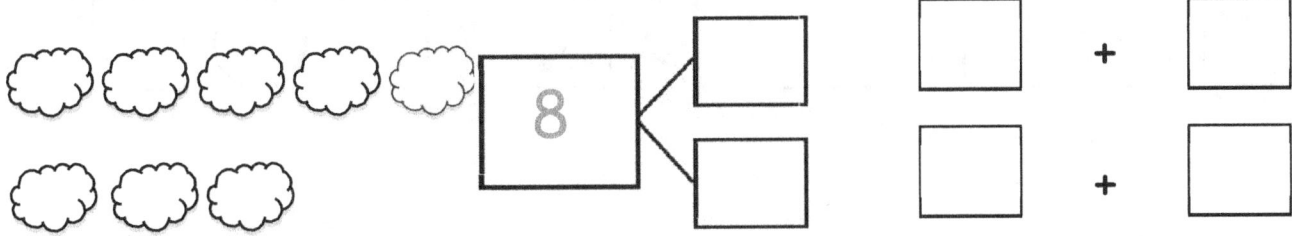

3. Circle 4. How many more does 4 need to make 8?

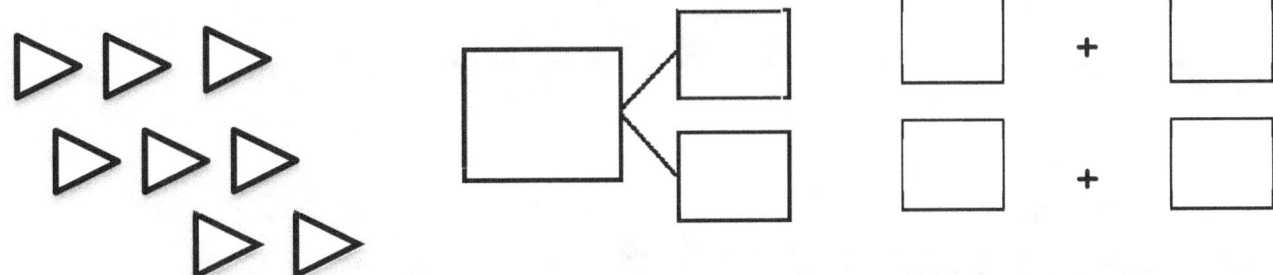

Fill in the missing part of the number bond and count on to find the total. Then write 2 addition sentences for each number bond.

1.    2.

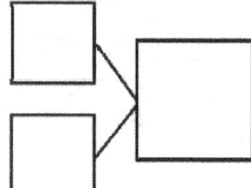

Shake Those Disks! - 8

8 0 8	8 1 7	8 2 6	8 3 5	8 4 4

Number Bond Dash!

Directions: Do as many as you can in 90 seconds. Write the amount you finished here: _____

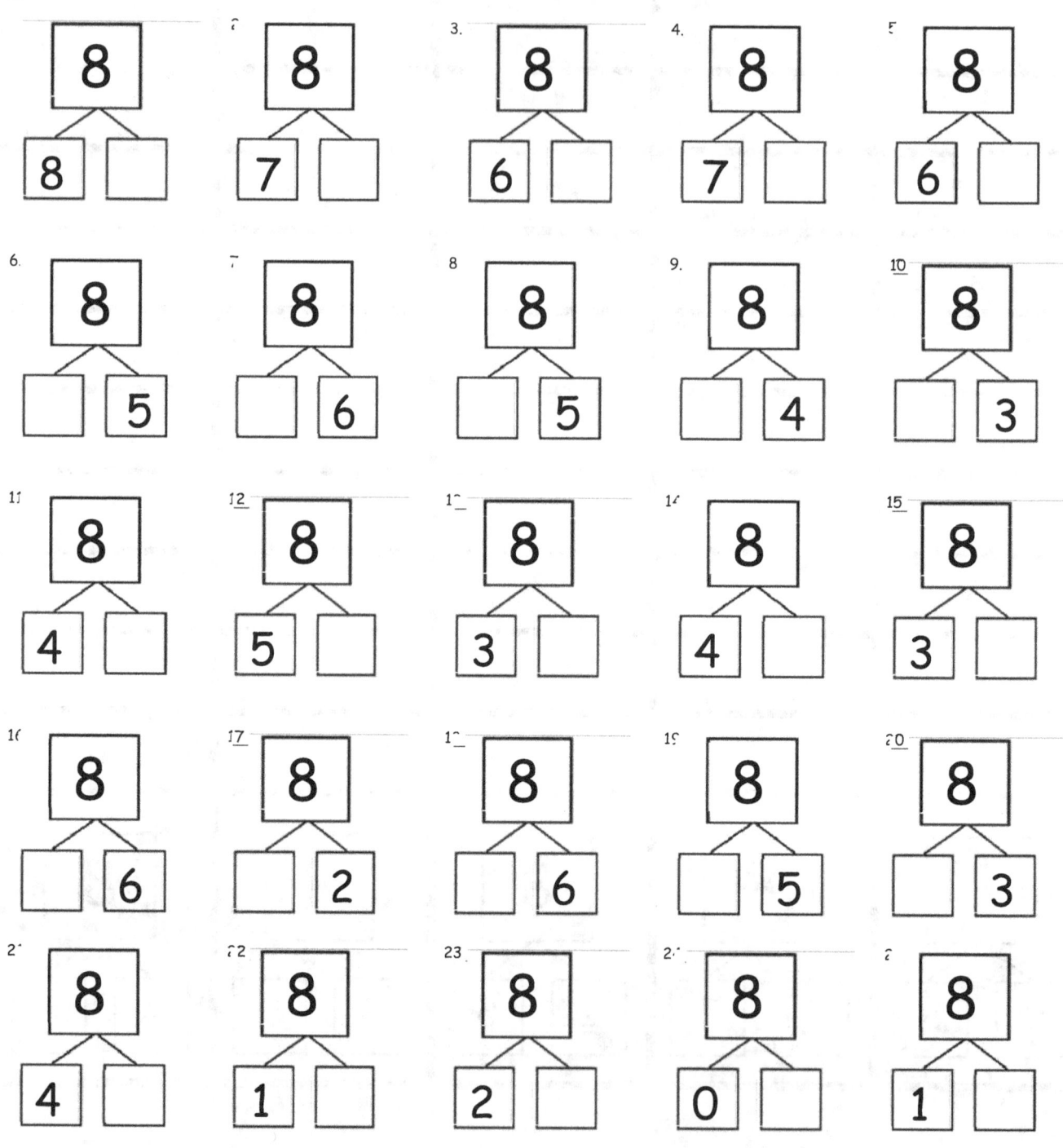

Circle the part. Count on to show 9 with the picture and number bond. Write the expressions.

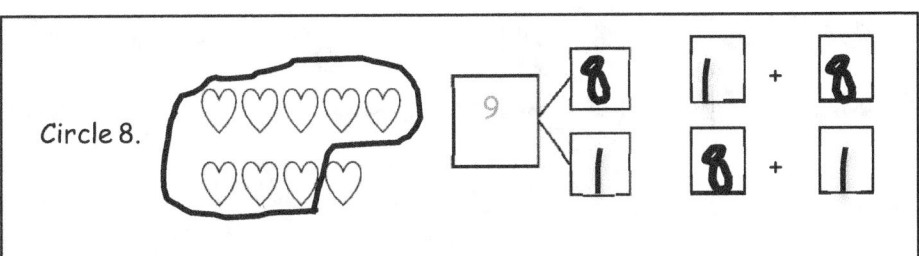

Circle 8.

1. Circle 7. How many more does 7 need to make 9?

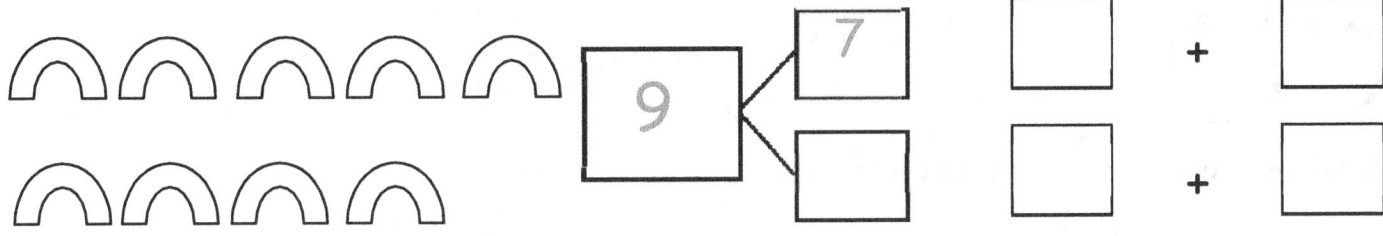

2. Circle 4. How many more does 4 need to make 9?

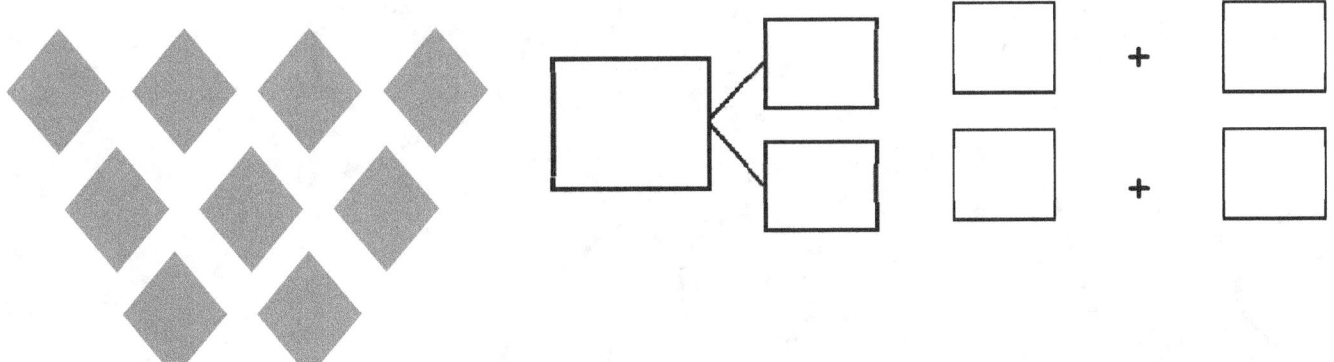

3. Circle 2. How many more does 2 need to make 9?

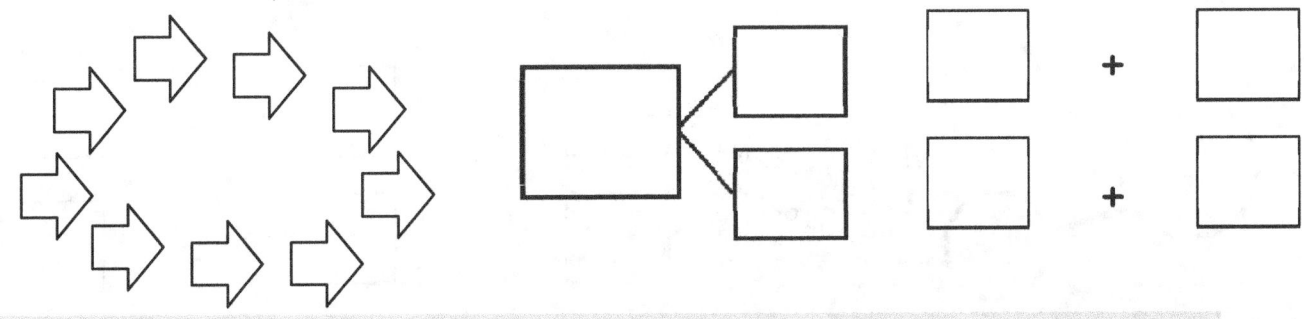

4. Draw a line to show partners of 9.

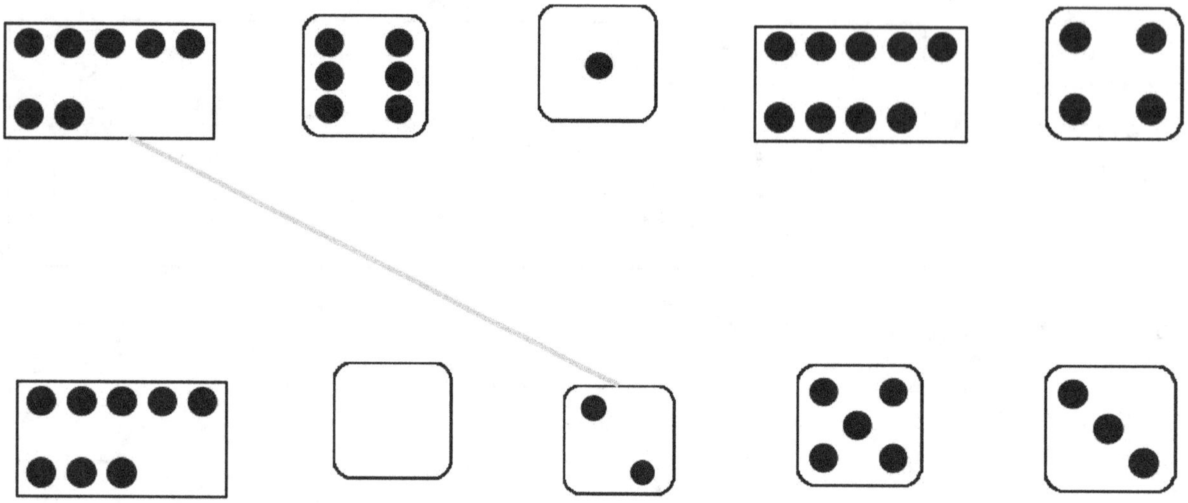

5. Write a number bond for each partner of 9. Use the partners above for help.

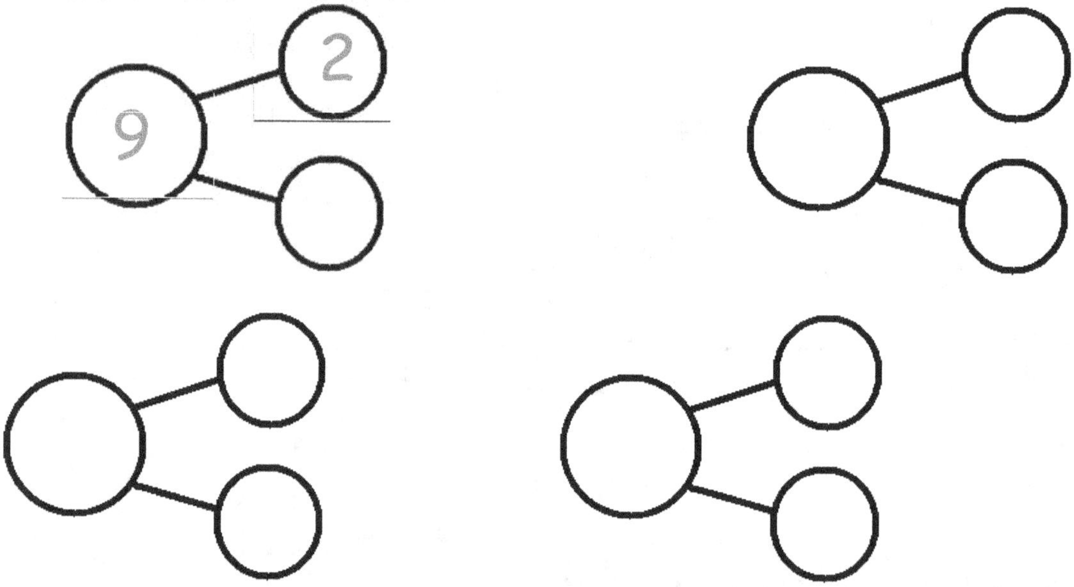

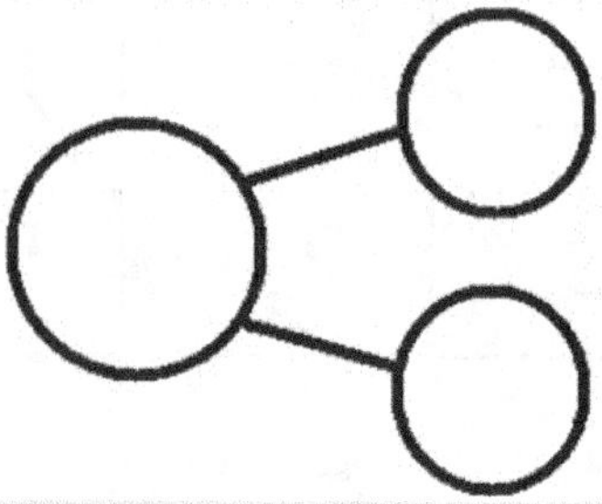

9 — 2

Write number sentences to match this number bond!

☐ + ☐ = ☐

☐ + ☐ = ☐

1. Circle the pairs of numbers that make 9.

2. Complete the number bonds and show 2 different ways to make 9.

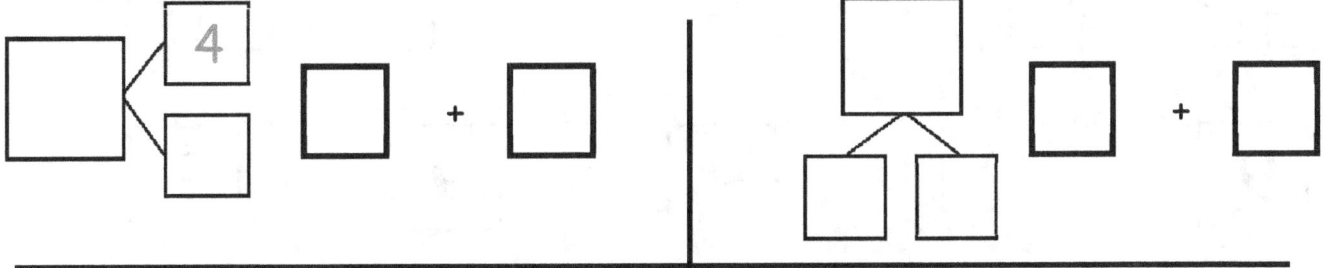

Number Bond Dash!

Directions: Do as many as you can in 90 seconds. Write the amount you finished here: _____

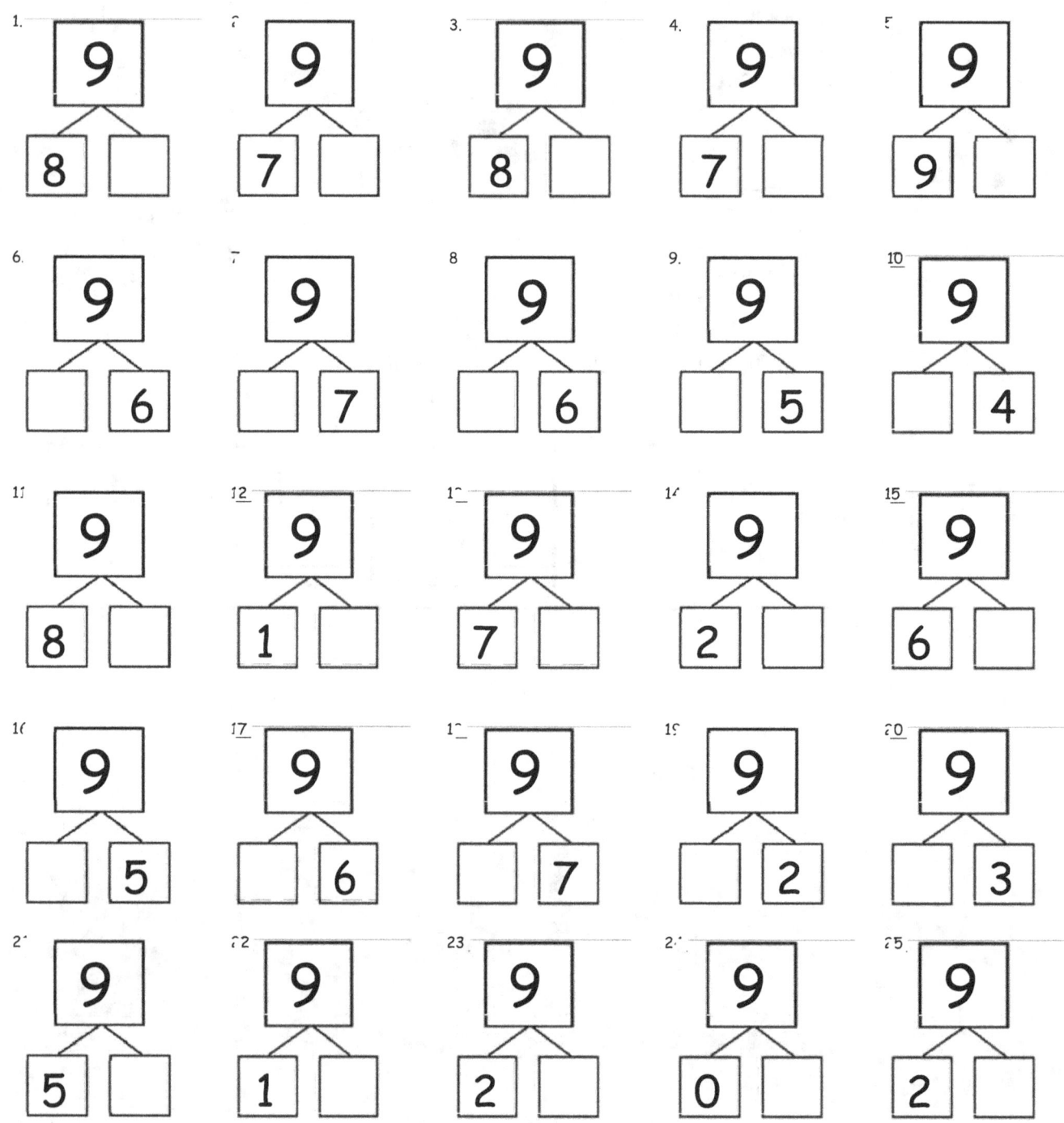

1. Use your bracelet to show different partners of 10. Then draw the beads. Write an expression to match.

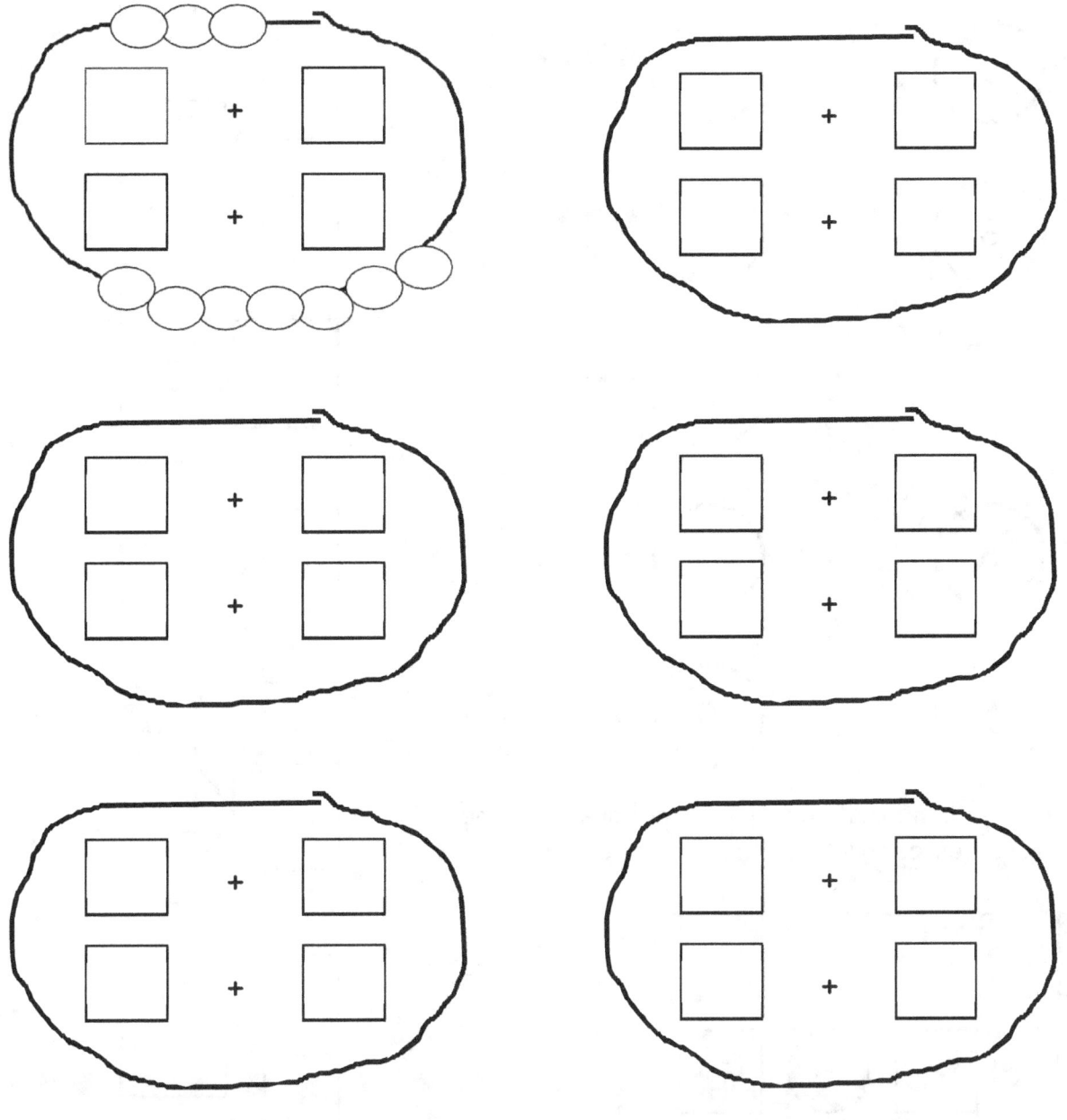

2. Match the partners of 10. Then write a number bond for each partner.

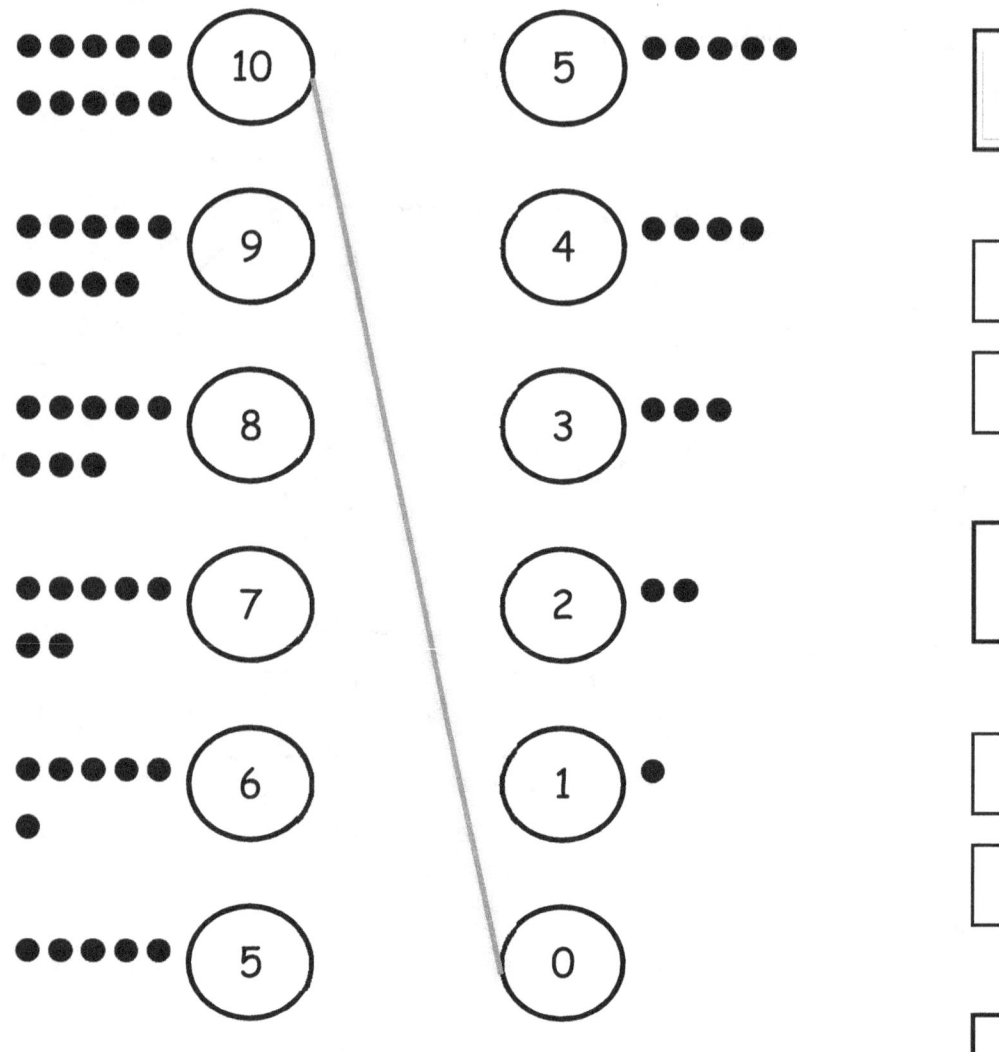

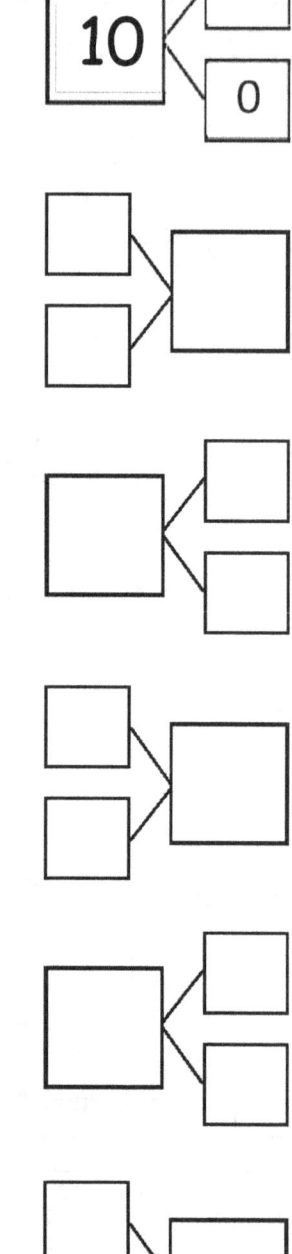

3. Color the number bond that has 2 parts that are the same.
Write addition sentences to match that number bond.

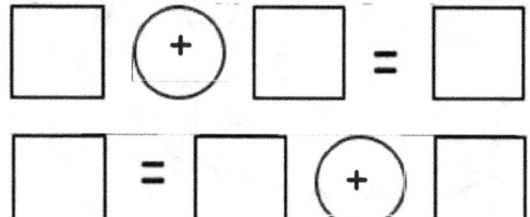

1. Color the partners that make 10.

Number Bond Dash!

Directions: Do as many as you can in 90 seconds. Write the amount you finished here: _____

1. 10 → 10, ___
2. 10 → 9, ___
3. 10 → 8, ___
4. 10 → 9, ___
5. 10 → 10, ___
6. 10 → ___, 9
7. 10 → ___, 8
8. 10 → ___, 7
9. 10 → ___, 8
10. 10 → ___, 7
11. 10 → 6, ___
12. 10 → 7, ___
13. 10 → 6, ___
14. 10 → 5, ___
15. 10 → 4, ___
16. 10 → ___, 6
17. 10 → ___, 4
18. 10 → ___, 3
19. 10 → ___, 4
20. 10 → ___, 3
21. 10 → 0, ___
22. 10 → 1, ___
23. 10 → 2, ___
24. 10 → 4, ___
25. 10 → 2, ___

1.

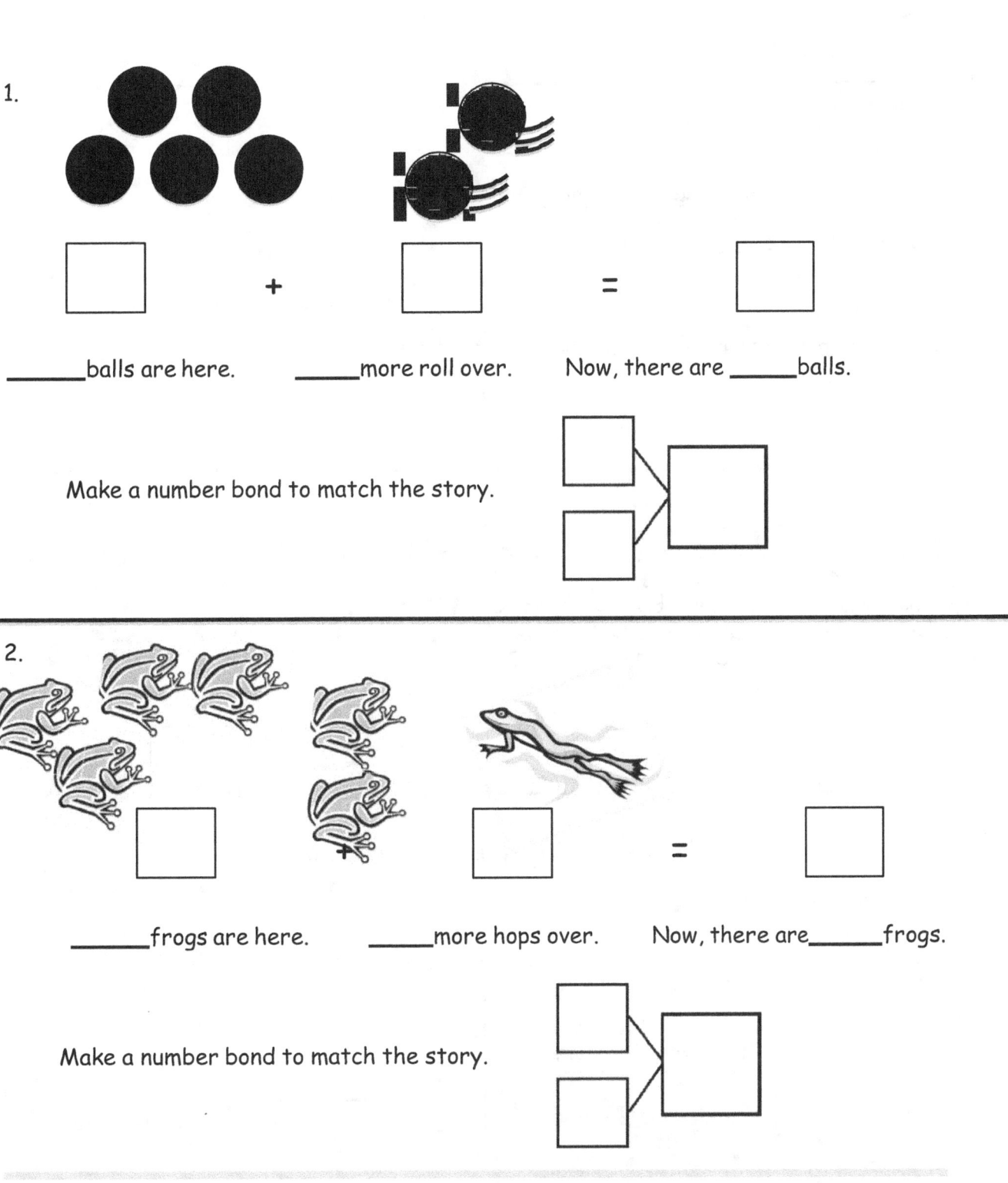

☐ + ☐ = ☐

_____ balls are here. _____ more roll over. Now, there are _____ balls.

Make a number bond to match the story.

2.

☐ + ☐ = ☐

_____ frogs are here. _____ more hops over. Now, there are _____ frogs.

Make a number bond to match the story.

3.

☐ + ☐ = ☐

There are _____ dark flags.   There are _____ white flags.

Altogether, there are _____ flags.

Make a number bond to match the story.

4.

☐ + ☐ = ☐

There are _____ white flowers.   There are _____ dark flowers.

Altogether, there are _____ flowers.

Make a number bond to match the story.

Draw a picture and write a number sentence to match the story.

1. Ben has 3 red balls 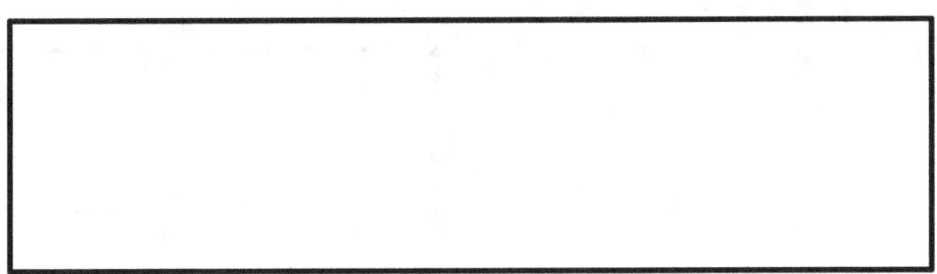 and gets 5 green balls. How many balls does he have now?

☐ + ☐ = ☐   Ben has _____ balls.

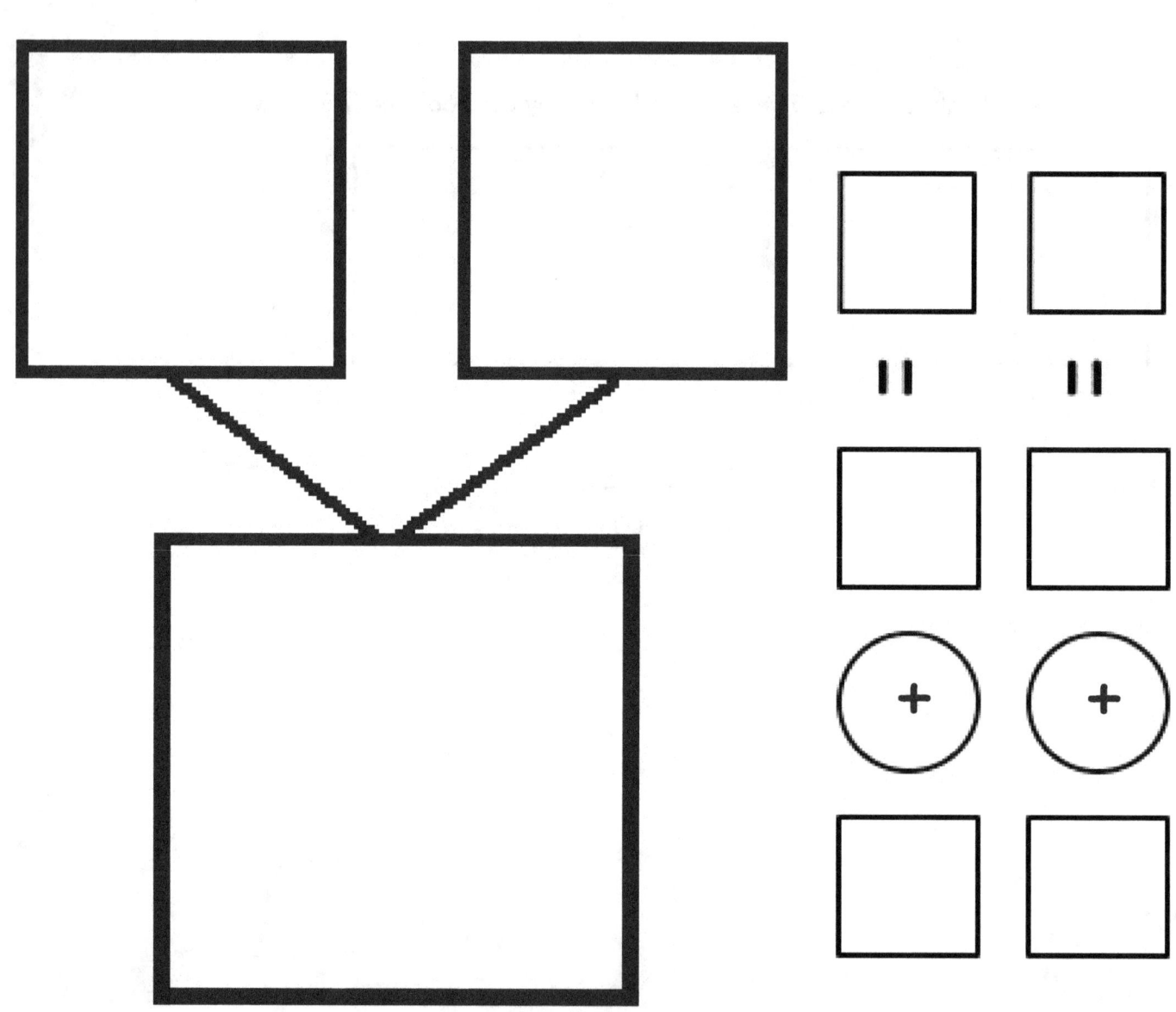

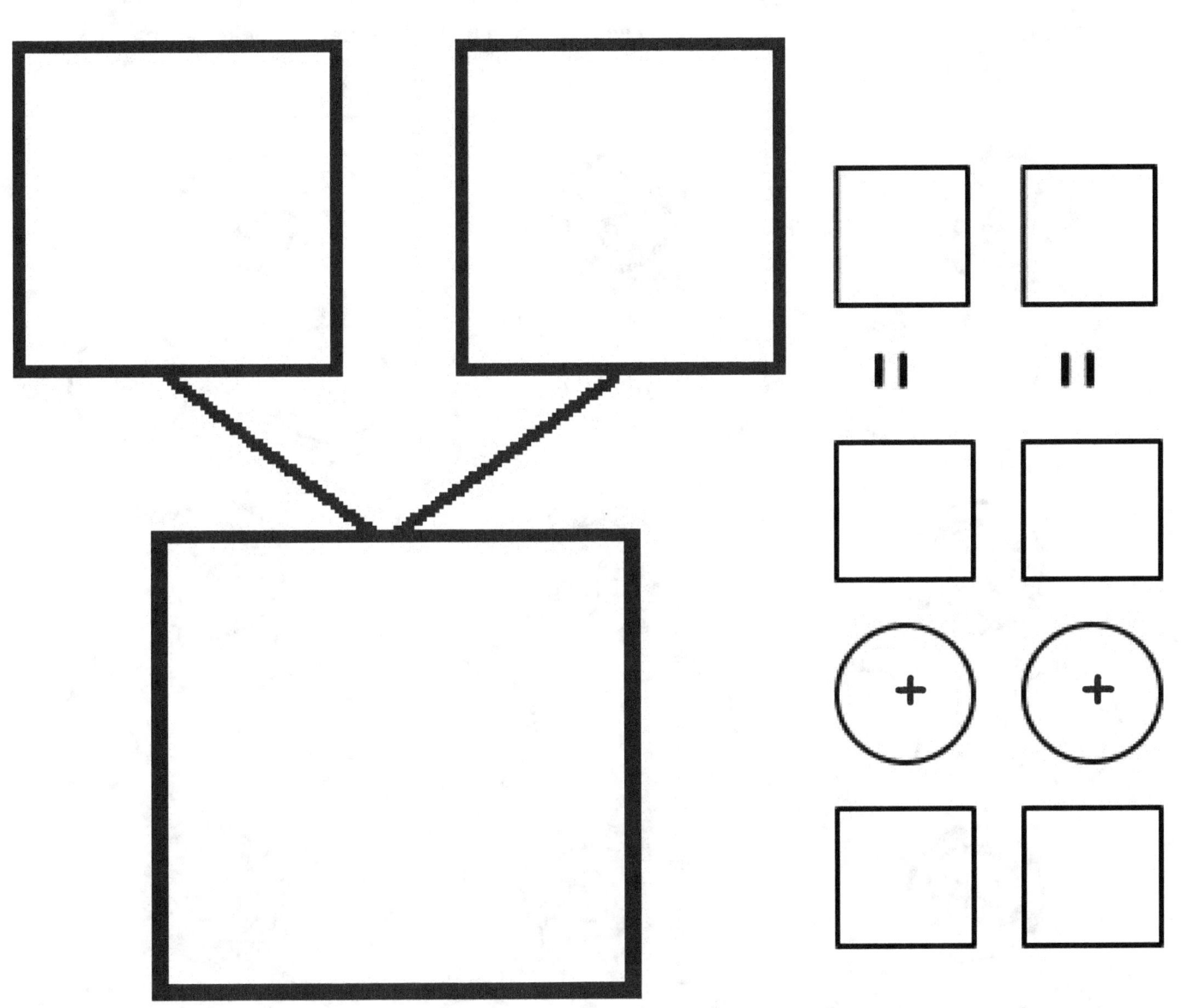

Target Practice

Directions: Choose a *target number* between 6 and 10 and write it in the middle of the circle on the top of the page. Roll a die. Write the number rolled in the circle at the end one of the arrows. Then, make a bull's-eye by writing the number needed to make your target in the other circle.

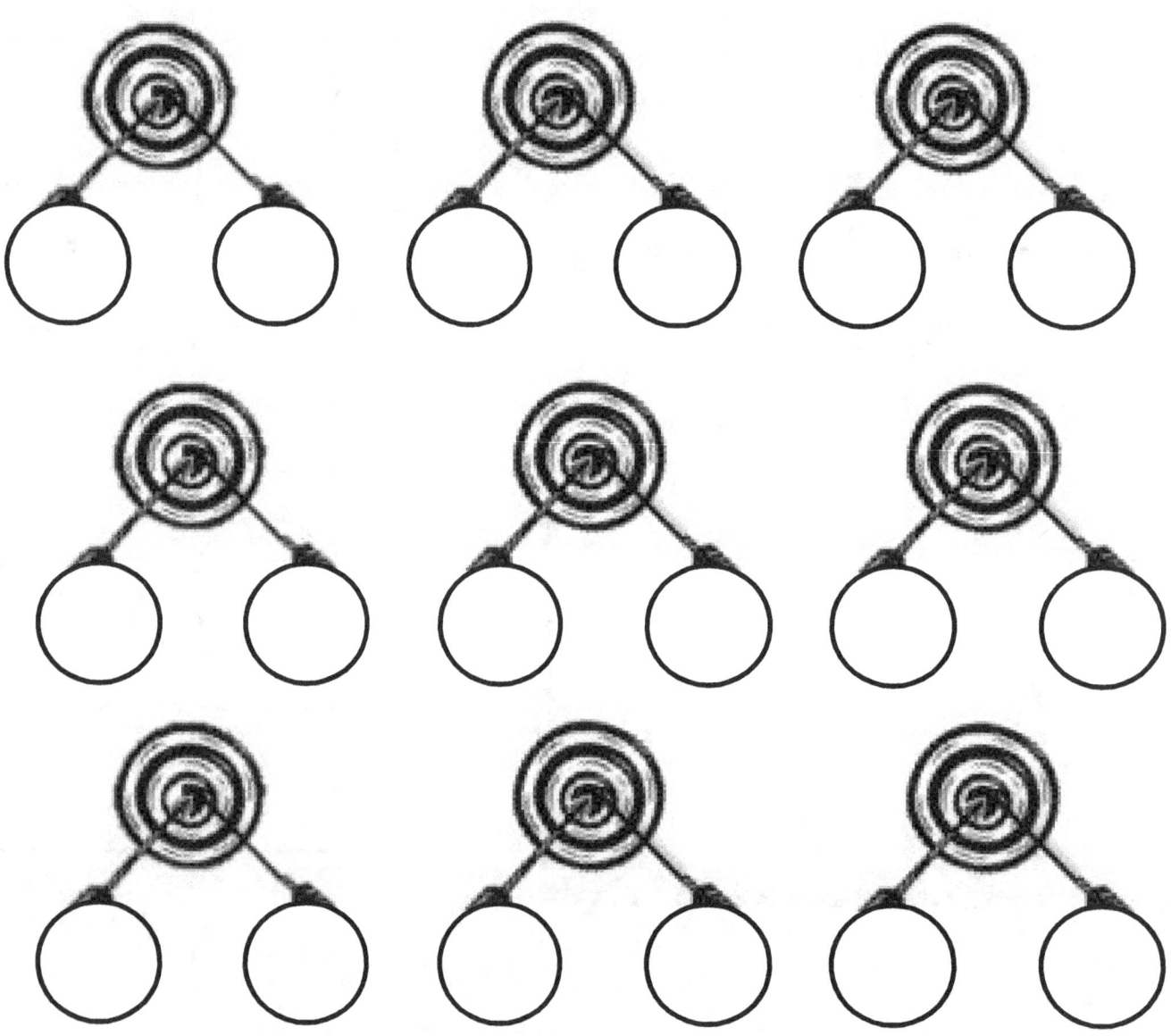

1. Use the picture to write the number sentence and the number bond.

_____little turtles   +   _____big turtles   = _____ turtles

2.

_____dogs that are awake   +   _____sleeping dogs   = _____ dogs

3.

_____pigs   +   _____pigs in mud   =_____pigs

4. Draw a line from the picture to the matching 5-group cards.

Draw to show the story. There are 3 large balls and 4 small balls.

☐ + ☐ = ☐

How many balls are there?   There are _____ balls.

Circle the set of numeral tiles that match your picture.

1. Jill was given a total of 5 flowers for her birthday. Draw more flowers in the vase to show Jill's birthday flowers.

How many flowers did you have to draw? ____ flowers

Write a number sentence and a number bond to match the story.

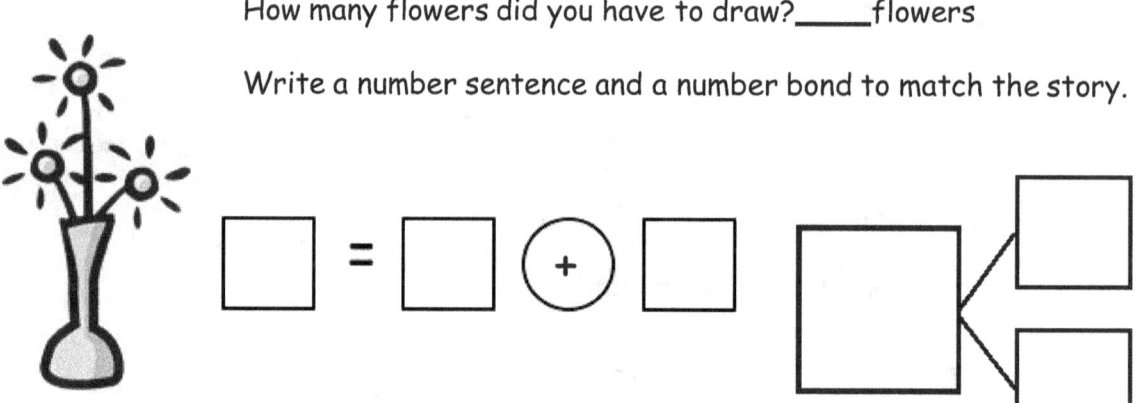

2. Kate and Nana were baking cookies. They made 2 heart cookies and then made some square cookies. They made 8 cookies altogether. How many square cookies did they make? Draw and count on to show the story.

Write a number sentence and a number bond to match the story.

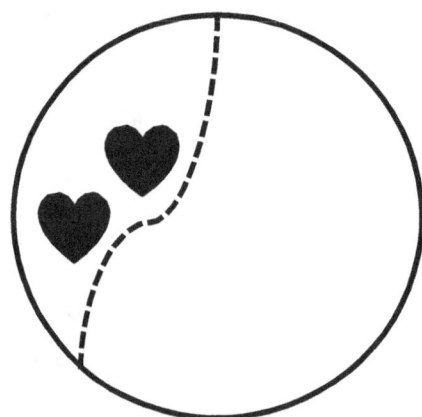

Show the parts. Write a number bond to match the story.

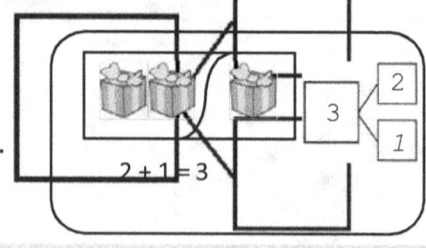

3. Bill has 2 trucks. His friend, James came over with some more. Together they had 5 trucks. How many trucks did James bring over?

 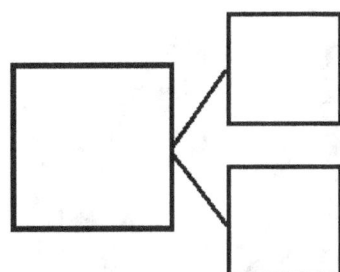

James brought over _____ trucks.

Write a number sentence to explain the story.

2 + ☐ = 5

4. Jane caught 7 fish before she stopped to eat lunch. After lunch she caught some more . At the end of the day she had 9 fish. How many fish did she catch after lunch?

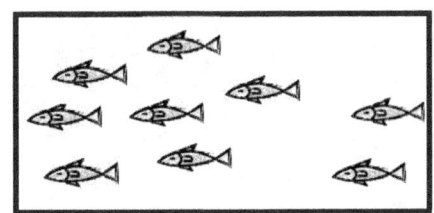 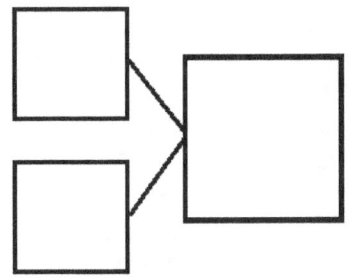

Jane caught _____ fish after lunch.

Write a number sentence to explain the story.

☐ + ☐ = ☐

Name _____ Date _____

1. Draw more bears to show that Jen has 8 bears total.

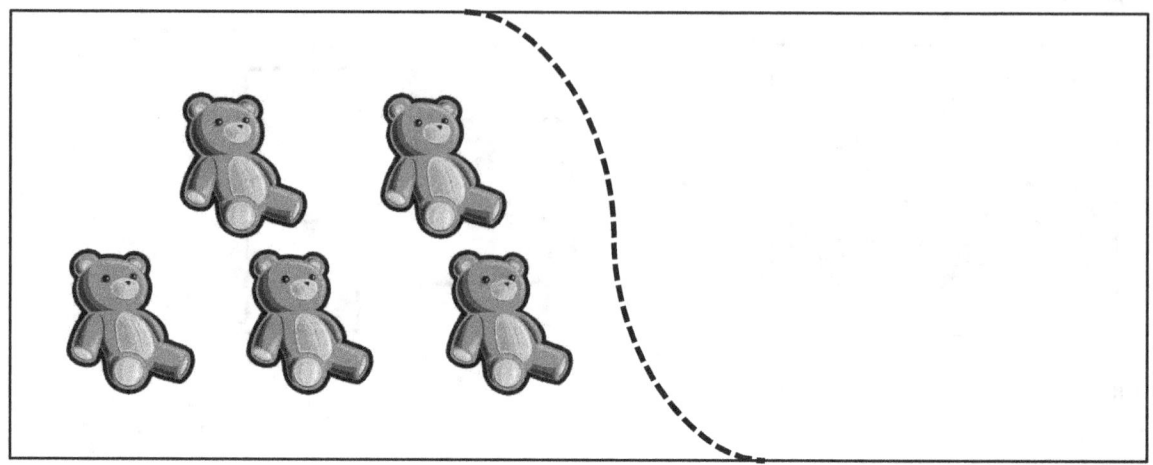

I added _____ more bears.

Write a number sentence to show how many bears you drew.

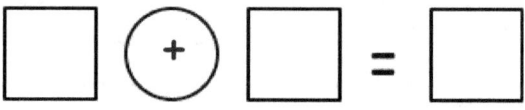

Number Sentence Cards

3	+	2	=	5
7	+	1	=	8
6	+	1	=	7
4	+	2	=	6
6	=	5	+	1
10	=	7	+	3
8	=	6	+	2
7	=	5	+	2

NYS COMMON CORE MATHEMATICS CURRICULUM　　　Lesson 12 Problem Set　1•1

 Fill in the missing numbers.

1.

   3   +   ___   =   5

2.

   5   +   ___   =   9

3.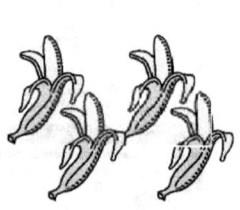

   4   +   ___   =   10

4. Kate and Bob had 6 balls at the park. Kate had 2 of the balls.

How many balls did Bob have?

_____ balls = _____ balls + _____ balls

Bob had _____ balls at the park.

---

5. I had 3 apples. My mom gave me some more. Then I had 10 apples.

How many apples did my mom give me?

_____ apples + _____ apples = _____ apples

Mom gave me _____ apples.

NYS COMMON CORE MATHEMATICS CURRICULUM — Lesson 12 Exit Ticket — 1•1

Draw a picture and count on to solve the math story.

Bob caught 5 fish. John caught some more fish. They had 7 fish in all. How many fish did John catch?

Write a number sentence to match your picture.

☐ = ☐ + ☐

John caught _____ fish.

With a partner, create a story for each of the number sentences below. Draw a picture to show. Write the number bond to match the story.

1. 6 + 2 = ☐

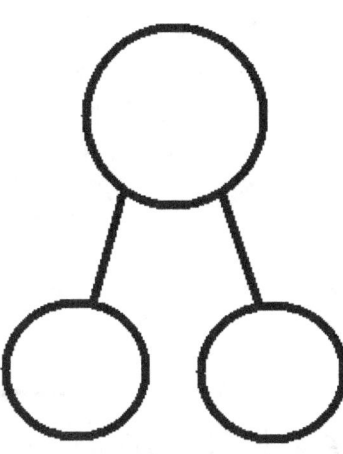

2. 5 + 5 = ☐

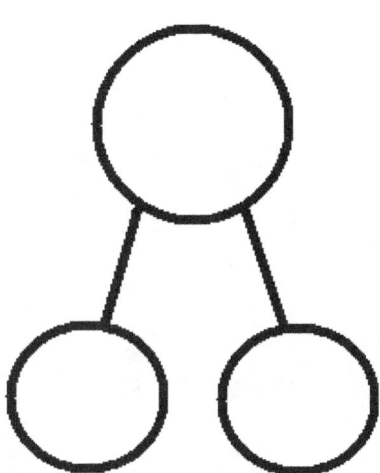

3. 5 + ☐ = 7

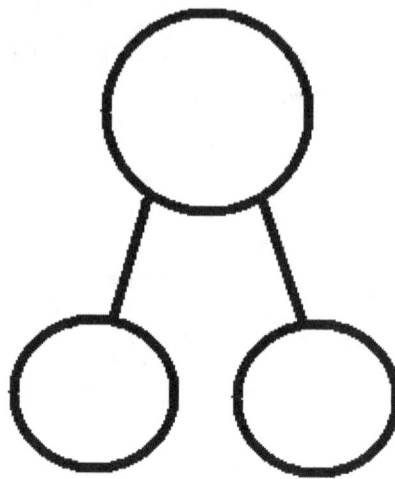

4. 6 + ☐ = 10

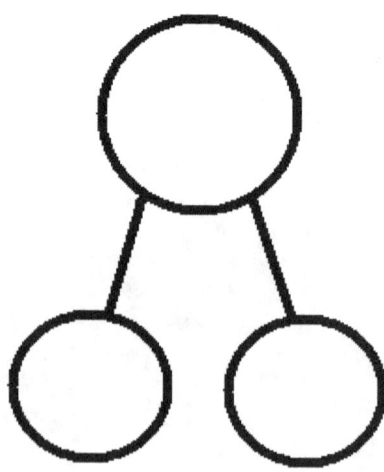

Tell a math story for each number sentence by drawing a picture.

1. 5 + 1 = 6

2. 3 + ? = 8

1. Count on to add.

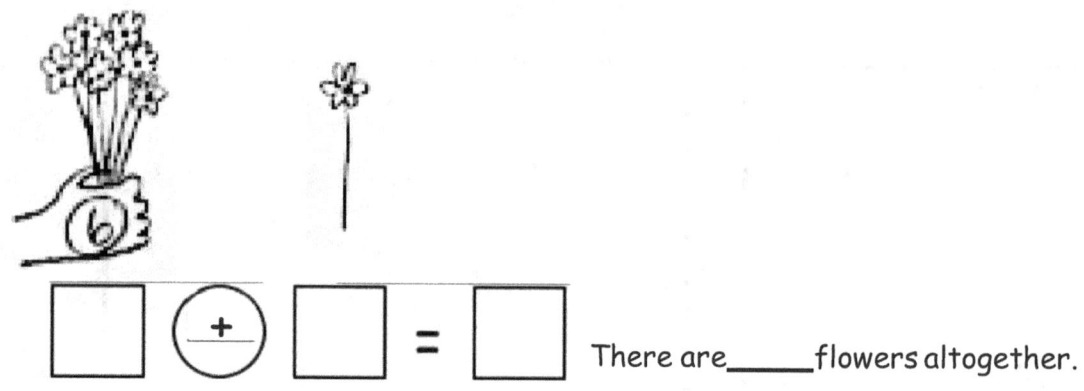

☐ ⊕ ☐ = ☐   There are ____ flowers altogether.

2.

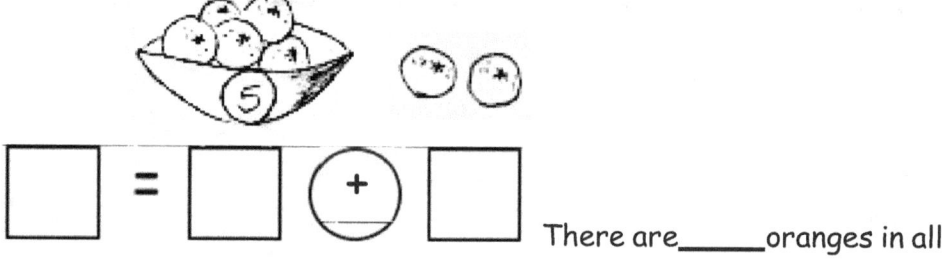

☐ = ☐ ⊕ ☐   There are ____ oranges in all.

3.

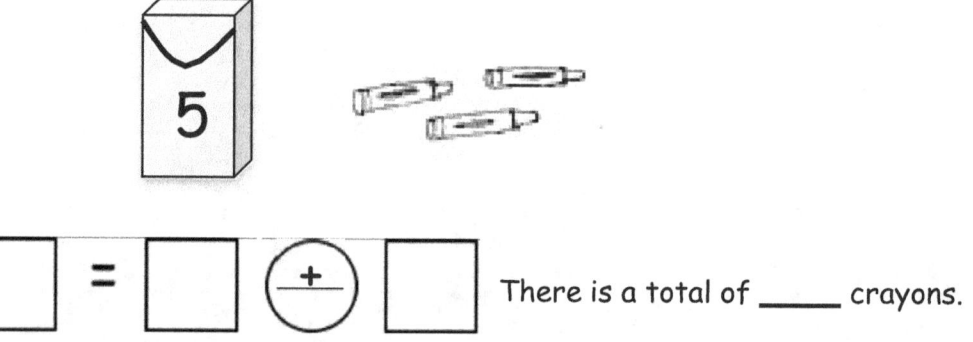

☐ = ☐ ⊕ ☐   There is a total of ____ crayons.

4. Use your 5-group cards to count on to add. Try to use as few dot cards as you can.

6 + 1 = ▢

6 + 3 = ▢

7 + 2 = ▢

▢ = 5 + 3

 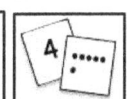

5. Use your 5-group cards, your fingers or your known facts to count on to add..

8 + 2 = ▢

▢ = 4 + 1

4 + 3 = ▢

▢ = 6 + 3

NYS COMMON CORE MATHEMATICS CURRICULUM　　　Lesson 14 Exit Ticket　1•1

 6 　  　　6 + 2 = ☐

I counted _____ more hats.

Count on to solve the number sentences.

7 + 3 = ☐ 　　　 8 + 2 = ☐

Lesson 15 Sprint 1•1

*Count on to add.

1	1 + 1		16	4 + 3
2	2 + 1		17	5 + 3
3	3 + 1		18	7 + 3
4	3 + 2		19	7 + 2
5	1 + 2		20	8 + 2
6	2 + 2		21	6 + 2
7	2 + 3		22	6 + 1
8	2 + 1		23	6 + 1
9	2 + 2		24	6 + 2
10	3 + 2		25	7 + 2
11	5 + 2		26	8 + 2
12	8 + 2		27	2 + 8
13	8 + 1		28	2 + 6
14	7 + 1		29	3 + 6
15	9 + 1		30	4 + 5

Count and write the number.

1	1 + 1		16	4 + 2	
2	2 + 2		17	3 + 2	
3	3 + 2		18	5 + 2	
4	2 + 2		19	7 + 2	
5	2 + 1		20	7 + 3	
6	3 + 1		21	6 + 3	
7	3 + 2		22	6 + 2	
8	3 + 2		23	6 + 2	
9	2 + 2		24	5 + 2	
10	4 + 2		25	7 + 2	
11	1 + 2		26	6 + 2	
12	2 + 1		27	2 + 6	
13	3 + 1		28	2 + 7	
14	5 + 1		29	3 + 7	
15	7 + 1		30	4 + 7	

1. Count on to add.

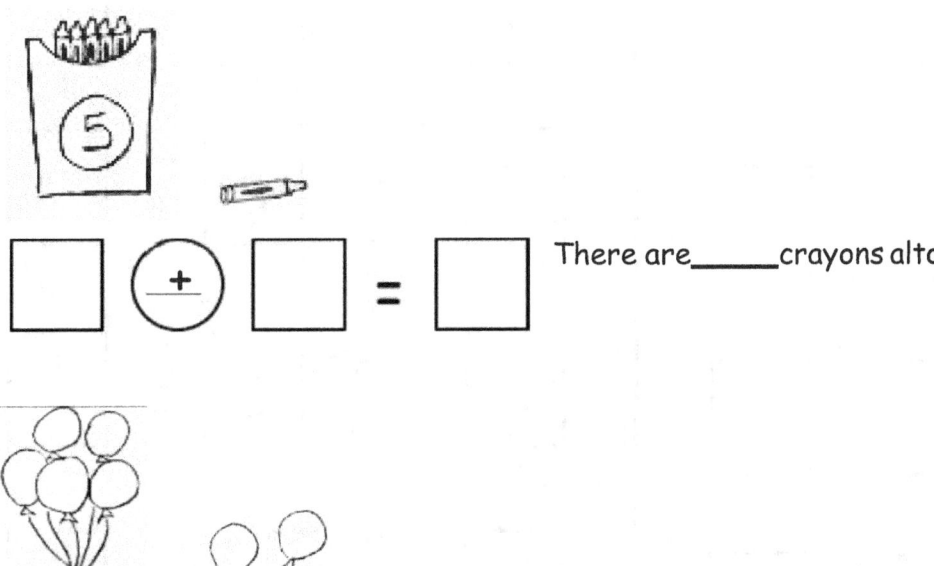

There are _____ crayons altogether.

There are a total of _____ balloons.

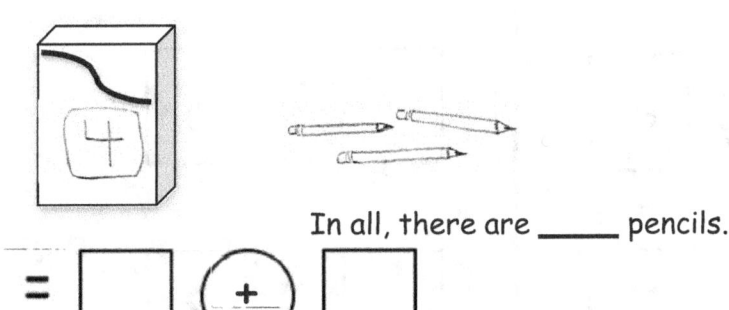

In all, there are _____ pencils.

2. What shortcut or efficient strategy can you find to add?

4 + 1 = ☐   2 + 5 = ☐

4 + 3 = ☐   7 + 2 = ☐

7 + 1 = ☐   7 + 3 = ☐

☐ = 6 + 2   ☐ = 4 + 2

☐ = 5 + 3   ☐ = 2 + 5

☐ = 3 + 6   ☐ = 6 + 2

☐ = 3 + 7   ☐ = 2 + 8

NYS COMMON CORE MATHEMATICS CURRICULUM Lesson 15 Exit Ticket 1•1

Use the picture to add.   Show the shortcut you used to add.

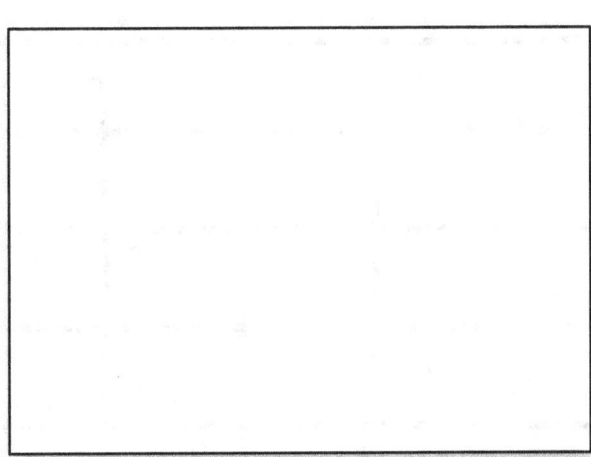

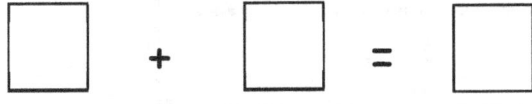

There are _____ eggs total.

Shake Those Disks! - 7

7 / 0 7	7 / 1 6	7 / 2 5	7 / 3 4

© Kelly Spinks

1. Draw more apples to solve 4 + ? = 6.

I added _____ apples to the tree.

2. How many more to make 7?

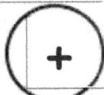

3. How many more to make 8?

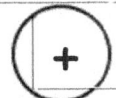

4. How many more to make 9?

NYS COMMON CORE MATHEMATICS CURRICULUM   Lesson 16 Problem Set   1•1

$3 + 1 = 4$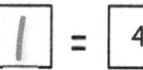

5. Count on to add. Circle the strategy you used to keep track.

$4 + \square = 5$

$4 + \square = 7$

$8 = 5 + \square$

$10 = \square + 8$

$7 + \square = 8$

$\square + 5 = 7$

$8 = 6 + \square$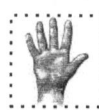

$10 = \square + 7$

Solve the number sentences. Circle the tool or strategy you used.

5 + ☐ = 7   I counted on _____ using

Or

I just knew

---

6 + ☐ = 9   I counted on _____ using

Or

I just knew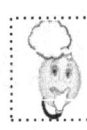

NYS COMMON CORE MATHEMATICS CURRICULUM   Lesson 17 Problem Set   1•1

Write an expression that matches the groups on each plate. If the plates have the same amount of fruit, write the equal sign between the expressions.

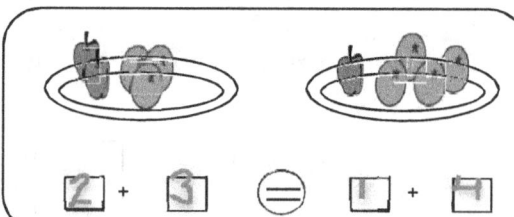

1.

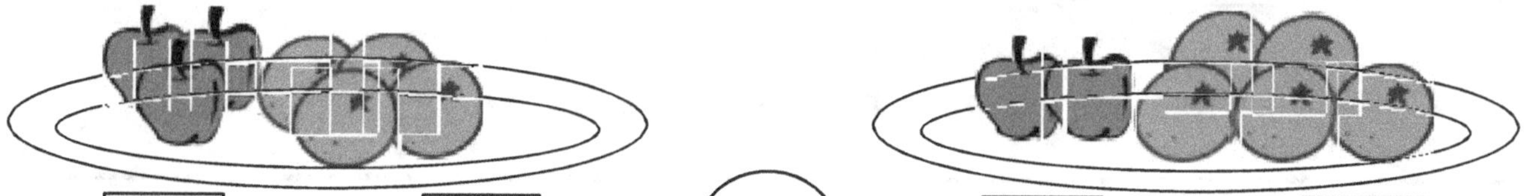

2.

3.

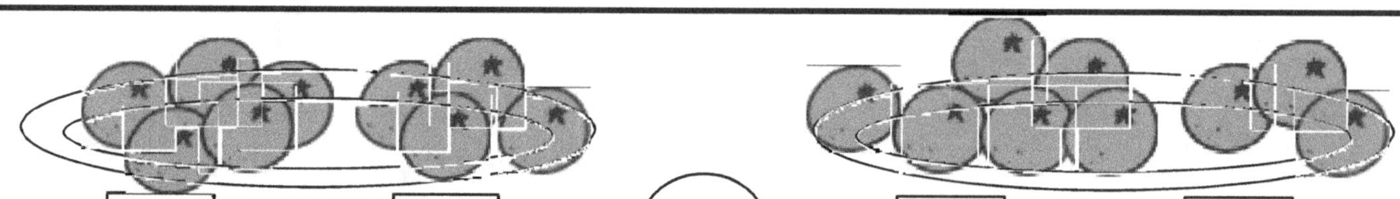

4.

NYS COMMON CORE MATHEMATICS CURRICULUM   Lesson 17 Problem Set   1•1

5. Write an expression to match each domino.

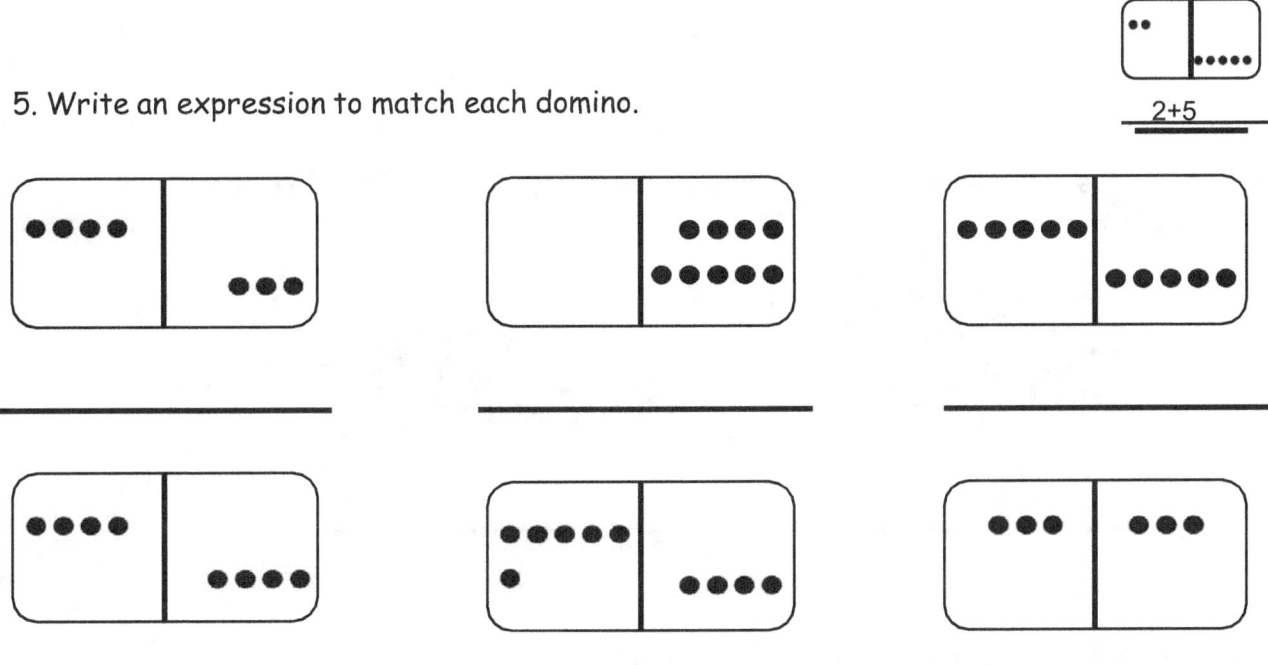

Find two sets of expressions that are equal. Connect them below with = to make true number sentences.

6.

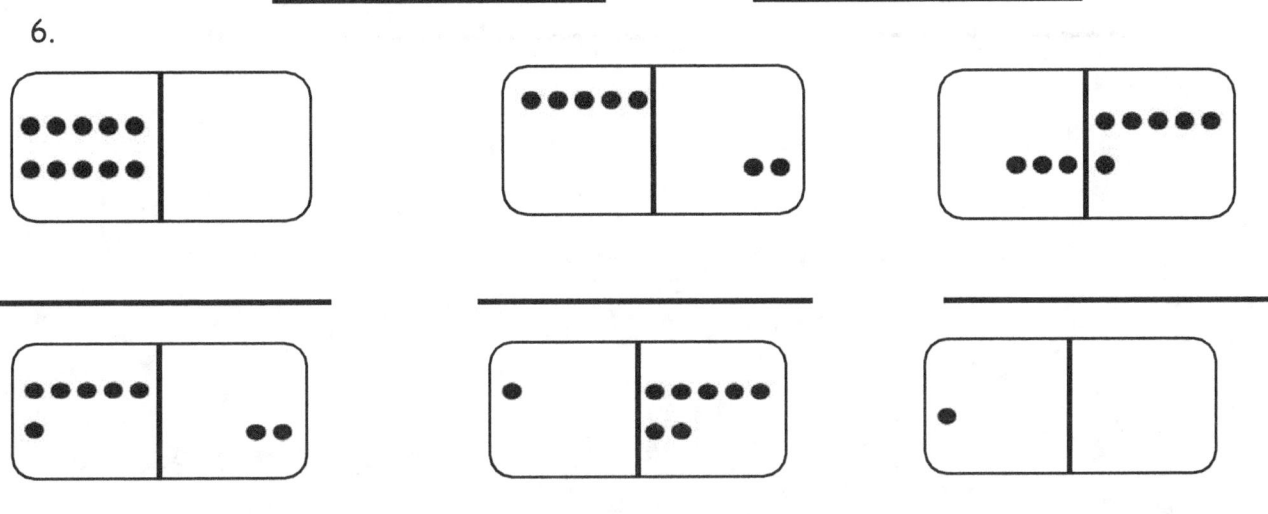

Find two sets of expressions that are equal. Connect them below with = to make true number sentences.

NYS COMMON CORE MATHEMATICS CURRICULUM   Lesson 17 Exit Ticket   1•1

Use math drawings to make the pictures equal. Connect them below with = to make true number sentences.

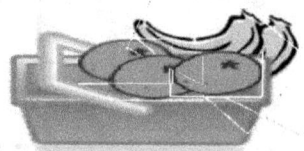   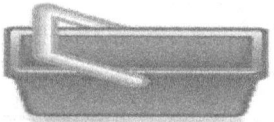

_____   _____

Shade the equal dominoes. Write a true number sentence.

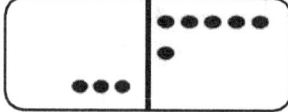   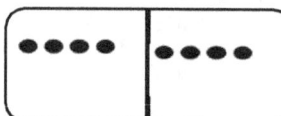   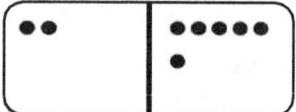

_____   _____

NYS COMMON CORE MATHEMATICS CURRICULUM — Lesson 18 Worksheet — 1•1

1. Add. Color the balloons that match the number in the boy's mind. Find expressions that are equal. Connect them below with = to make true number sentences.

_____

_____

NYS COMMON CORE MATHEMATICS CURRICULUM Lesson 18 Worksheet 1•1

2. Are these number sentences true?   if it is true.   if it is false.

If it's false, re-write the number sentence to make it true.

(a) 3 + 1 = 2 + 2

(b) 9 + 1 = 1 + 2

_____   _____

(c) 2 + 3 = 1 + 4

(d) 5 + 1 = 4 + 2

_____   _____

(e) 4 + 3 = 3 + 5

(f) 0 + 10 = 2 + 8

_____   _____

(g) 6 + 3 = 4 + 5

(h) 3 + 7 = 2 + 6

_____   _____

3. Write a number in the expression and solve.   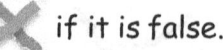 if it is true. if it is false.

1 + ____ = 3 + 2   ☐   ____ + 4 = 2 + 5   ☐

____ + 5 = 6 + ____   ☐   7 + ____ = 8 + ____

Find two ways to fix each number sentence to make it true.

✗ | 7 + 3 = 6 + 2 | | 8 + 1 = 3 + 5 |

___7 + 3___ = ___6 + 4___

_____ = _____   _____ = _____

_____ = _____   _____ = _____

True and False Number Sentence Cards

$4 + 1 = 2 + 2$	$2 + 5 = 8 + 2$
$3 + 2 = 4 + 1$	$9 + 1 = 4 + 6$
$6 + 2 = 3 + 3$	$3 + 4 = 6 + 3$
$1 + 7 = 4 + 4$	$5 + 4 = 3 + 7$
$2 + 5 = 4 + 3$	$5 + 5 = 6 + 3$
$5 + 1 = 4 + 2$	$8 + 2 = 3 + 7$

*Count On to Add

1	1 + 1		16	4 + 3	
2	2 + 1		17	3 + 3	
3	3 + 1		18	4 + 3	
4	3 + 2		19	3 + 4	
5	2 + 2		20	2 + 4	
6	3 + 2		21	4 + 2	
7	2 + 2		22	5 + 2	
8	3 + 0		23	2 + 5	
9	3 + 1		24	2 + 6	
10	3 + 2		25	6 + 3	
11	5 + 2		26	3 + 6	
12	5 + 3		27	2 + 7	
13	5 + 2		28	3 + 7	
14	5 + 3		29	2 + 8	
15	6 + 3		30	3 + 6	

*Count On to Add.

1	2 + 1		16	4 + 3	
2	1 + 1		17	3 + 3	
3	2 + 1		18	2 + 3	
4	2 + 2		19	1 + 3	
5	3 + 2		20	0 + 3	
6	2 + 2		21	1 + 3	
7	3 + 2		22	2 + 5	
8	3 + 1		23	5 + 2	
9	5 + 1		24	2 + 6	
10	6 + 1		25	6 + 2	
11	6 + 2		26	3 + 6	
12	5 + 2		27	3 + 7	
13	6 + 2		28	2 + 7	
14	6 + 3		29	2 + 6	
15	5 + 3		30	3 + 6	

1. Write the number bond to match the picture. Then complete the number sentences.

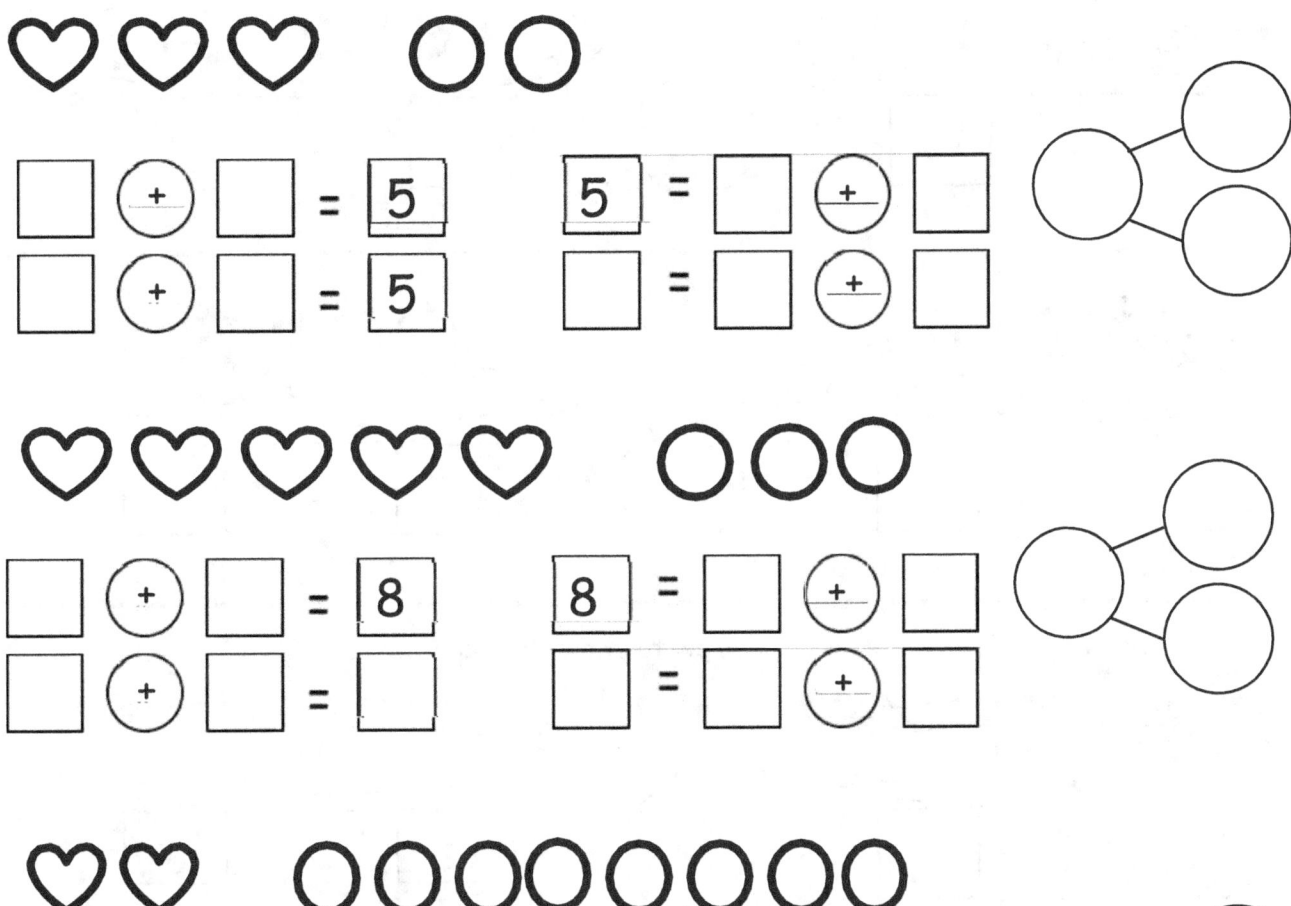

Write the expression under each plate. Add the equal sign to show they are the same amount

2.

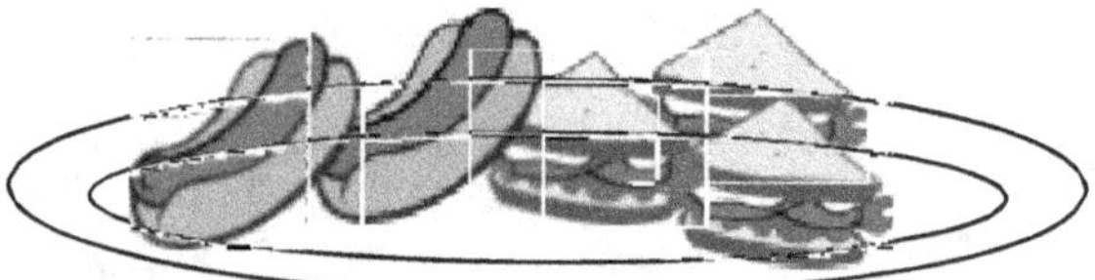

☐ + ☐ ◯ ☐ + ☐

3.

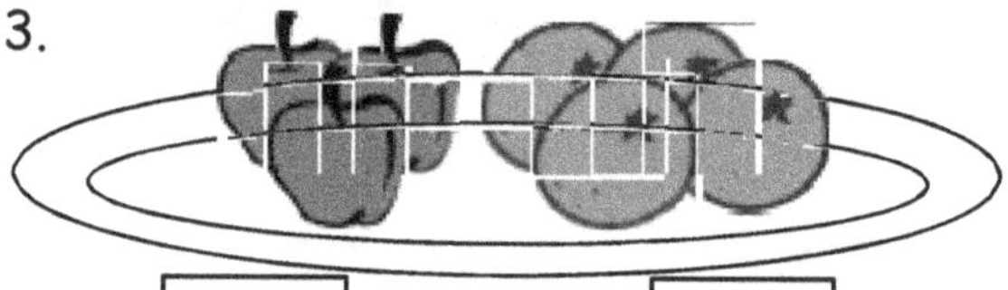

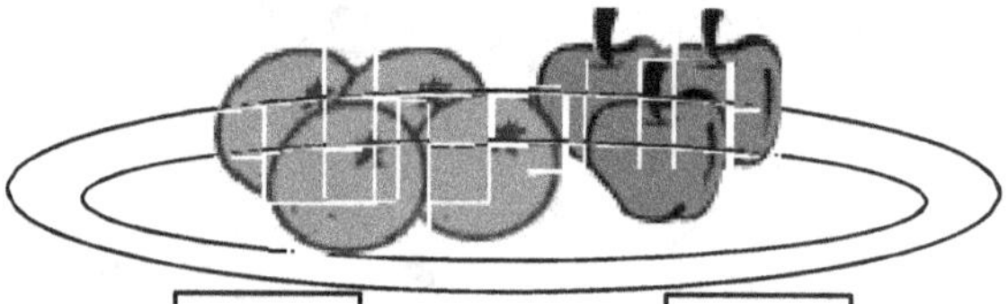

☐ + ☐ ◯ ☐ + ☐

4.

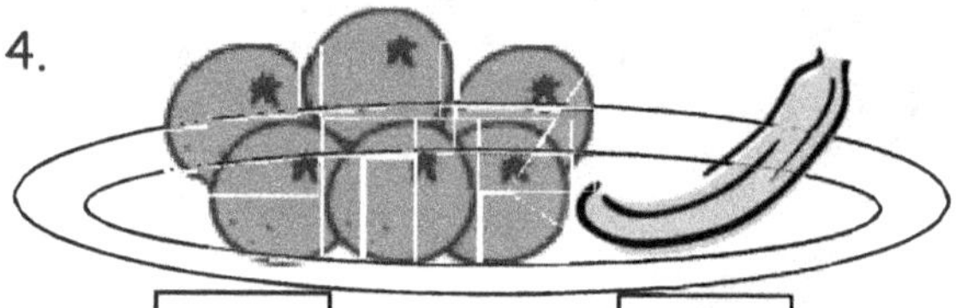

☐ + ☐ ◯ Draw to show the expression.

[1] + [6]

5. Draw and write to show 2 expressions that use the same numbers and have the same total.

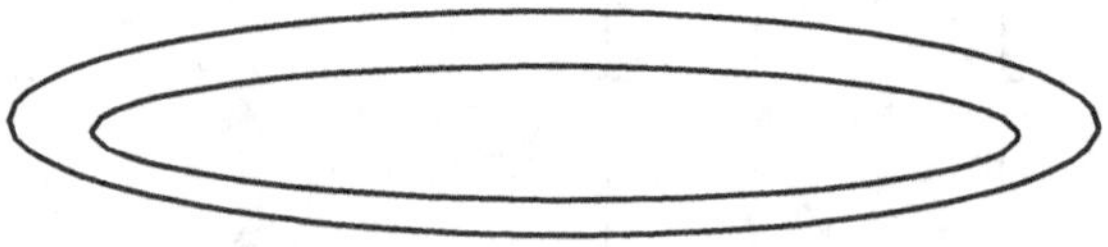

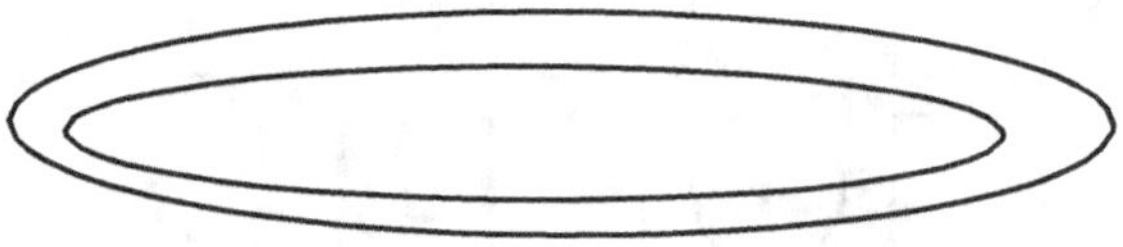

☐ + ☐ ◯ ☐ + ☐

Draw a picture and write the number sentences to show the parts in a different order.

___ + ___ = ___   ___ = ___ + ___

___ + ___ = ___   ___ = ___ + ___

Circle the larger amount and count on. Write the number sentence starting with the larger number.

1.

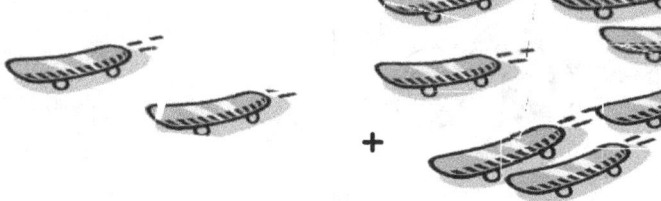

+

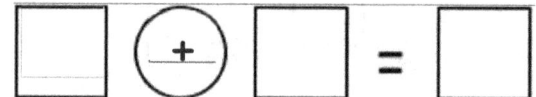

Color the larger part in the number bond. Write the number sentence starting with the larger number.

3 + 1 = 4

2.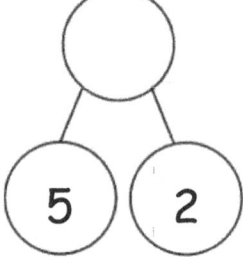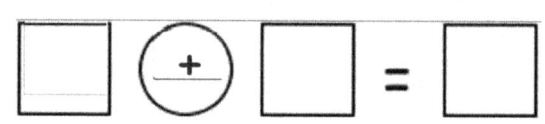

3.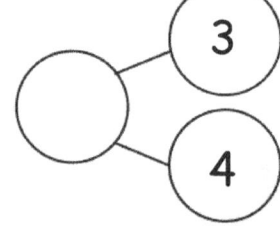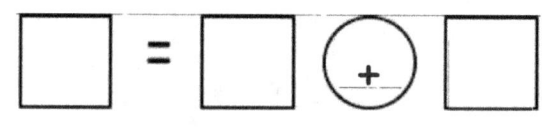

4.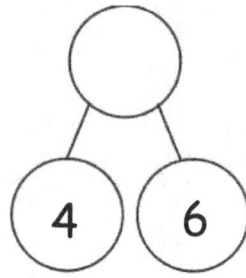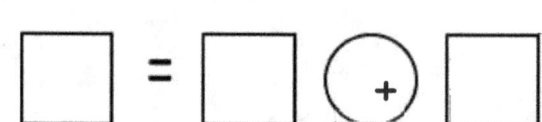

NYS COMMON CORE MATHEMATICS CURRICULUM   Lesson 20 Problem Set   1•1

Shade in the larger part of the bond. Count on from that part to find the total.
Rewrite the number sentence to start with the larger number.

5.

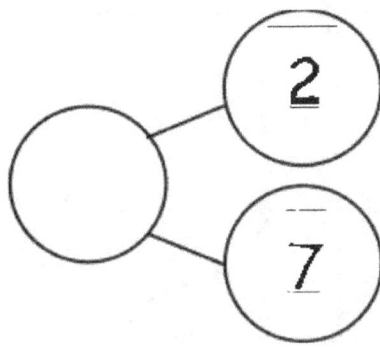

2 + ☐ = ☐

☐ + ☐ = ☐

6.

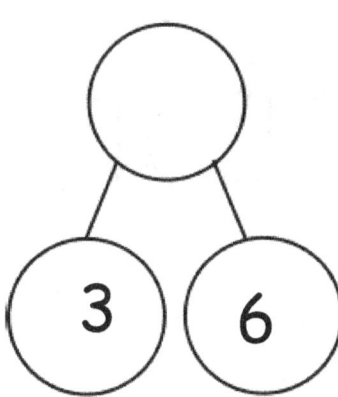

3 + ☐ = ☐

☐ + ☐ = ☐

Circle the larger number and count on to solve.

7. 1 + 5 = _____

8. 2 + 6 = _____

9. 4 + 3 = _____

10. 3 + 6 = _____

Circle the larger part, and complete the number bond. Write the number sentence starting with the larger part.

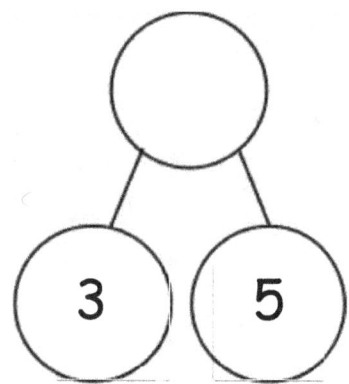 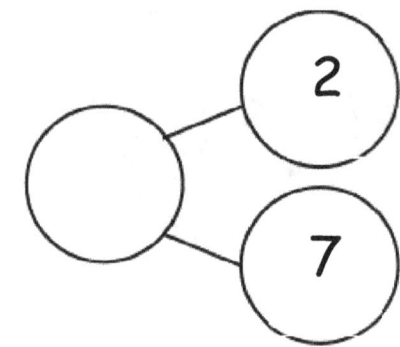

 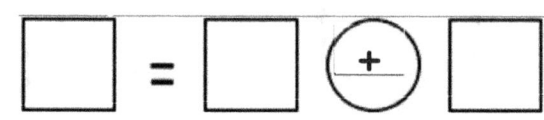

Add the numbers on the pairs of cards. Write the number sentences. Color doubles red. Color doubles plus 1 blue.

1.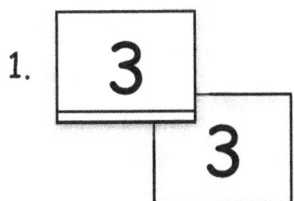

2.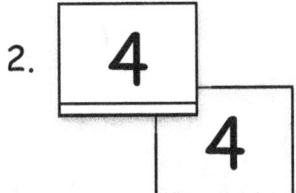

3.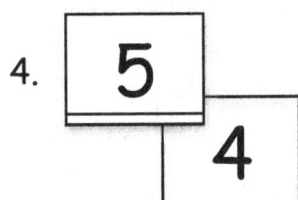

4.

Solve. Use your doubles to help. Draw and write the double that helped.

5. $5 + 4 = \square$ ○○○○○
                         ○○○○○

6. $4 + 3 = \square$ ○○○○○
                         ○○○○○

7. Solve the doubles and the doubles plus one number sentences.

(a) 0 + 0 = ☐   (a) 0 + 1 = ☐

(b) 2 + 2 = ☐   (b) 2 + 3 = ☐

(c) 3 + 3 = ☐   (c) 3 + 4 = ☐

(d) 4 + 4 = ☐   (d) 4 + 5 = ☐

(e) 3 + ☐ = 6   (e) 3 + ☐ = 7

(f) 5 + ☐ = 10   (f) 4 + ☐ = 9

8. Show how this strategy can help you solve: 5 + 6 = ☐

9. Write a set of 4 related addition facts for letter (*d*).

Write the double and double plus one number sentence for the 5- group card.

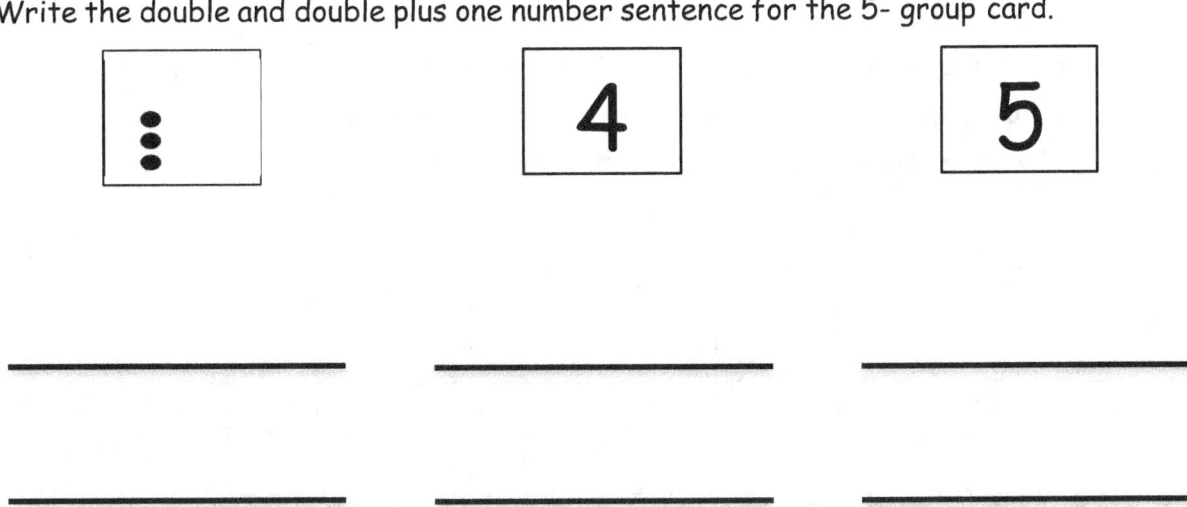

_____   _____   _____
_____   _____   _____

Use RED to color boxes with 0 as an addend. Find the total for each.
Use ORANGE to color boxes with 1 as an addend. Find the total for each.
Use YELLOW to color boxes with 2 as an addend. Find the total for each.
Use GREEN to color boxes with 3 as an addend. Find the total for each.
Use BLUE to color the boxes that are left. Find the total for each.

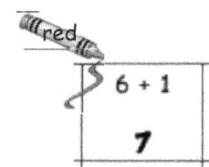

1 + 0	1 + 1	1 + 2	1 + 3	1 + 4	1 + 5	1 + 6	1 + 7	1 + 8	1 + 9
2 + 0	2 + 1	2 + 2	2 + 3	2 + 4	2 + 5	2 + 6	2 + 7	2 + 8	
3 + 0	3 + 1	3 + 2	3 + 3	3 + 4	3 + 5	3 + 6	3 + 7		
4 + 0	4 + 1	4 + 2	4 + 3	4 + 4	4 + 5	4 + 6			
5 + 0	5 + 1	5 + 2	5 + 3	5 + 4	5 + 5				
6 + 0	6 + 1	6 + 2	6 + 3	6 + 4					
7 + 0	7 + 1	7 + 2	7 + 3						
8 + 0	8 + 1	8 + 2							
9 + 0	9 + 1								

Some of the addends in this chart are missing! Fill in the missing numbers.

1+0	1+1	1+2	1+3	1+4	1+5	1+6	1+7	1+8	1+9
2+0	2+1	2+2	2+___	2+4	2+5	2+6	2+7	2+8	
3+0	3+1	3+2	3+___	3+4	3+5	3+6	3+7		
4+0	4+___	4+2	4+3	___+4	___+5	___+6			
5+0	5+___	5+2	5+3	5+4	5+5				
6+0	6+___	6+2	6+3	6+4					
7+___	7+1	7+2	7+3						
8+___	8+1	8+2							
9+___	9+1								
10+0									

Use your chart to write a list of number sentences in the spaces below.

Totals of 10	Totals of 9	Totals of 8	Totals of 7

Circle all the boxes that total 10. Make a straight line through all the boxes that total 8.

1 + 0	1 + 1	1 + 2	1 + 3	1 + 4	1 + 5	1 + 6	1 + 7	1 + 8	1 + 9
2 + 0	2 + 1	2 + 2	2 + 3	2 + 4	2 + 5	2 + 6	2 + 7	2 + 8	
3 + 0	3 + 1	3 + 2	3 + 3	3 + 4	3 + 5	3 + 6	3 + 7		
4 + 0	4 + 1	4 + 2	4 + 3	4 + 4	4 + 5	4 + 6			
5 + 0	5 + 1	5 + 2	5 + 3	5 + 4	5 + 5				
6 + 0	6 + 1	6 + 2	6 + 3	6 + 4					
7 + 0	7 + 1	7 + 2	7 + 3						
8 + 0	8 + 1	8 + 2							
9 + 0	9 + 1								

Friendly Fact Go Around: Addition Strategies Review

2 + 1 = ☐   3 + 1 = ☐   5 + 1 = ☐

4 + 1 = ☐   6 + 1 = ☐   9 + 1 = ☐

---

2 + 2 = ☐   2 + 3 = ☐   5 + 5 = ☐

3 + 3 = ☐   4 + 4 = ☐   4 + 5 = ☐

---

0 + 1 = ☐   1 + 3 = ☐   1 + 1 = ☐

2 + 2 = ☐   7 + 1 = ☐   3 + 3 = ☐

1 + 5 = ☐   5 + 5 = ☐   3 + 4 = ☐

8 + 1 = ☐   4 + 4 = ☐   5 + 4 = ☐

Solve the number sentences. Use the key to color. Once the box is colored, you do not need to color it again.

5 + 2 = ___	7 + 2 = ___	2 + 3 = ___
3 + 3 = ___	7 = 1 + ___	2 = 1 + ___
___ = 4 + 4	8 + 2 = ___	3 + 4 = ___
___ = 5 + 4	10 = 1 + ___	10 = 5 + ___

Color doubles - Red.

Color +1 - Blue

Color +2 - Green

Color doubles +1 - Brown

CHALLENGE:

List the number sentences that can be colored more than 1 way.

_____ _____

 Race to the Top!

2	4	6	0	8	10

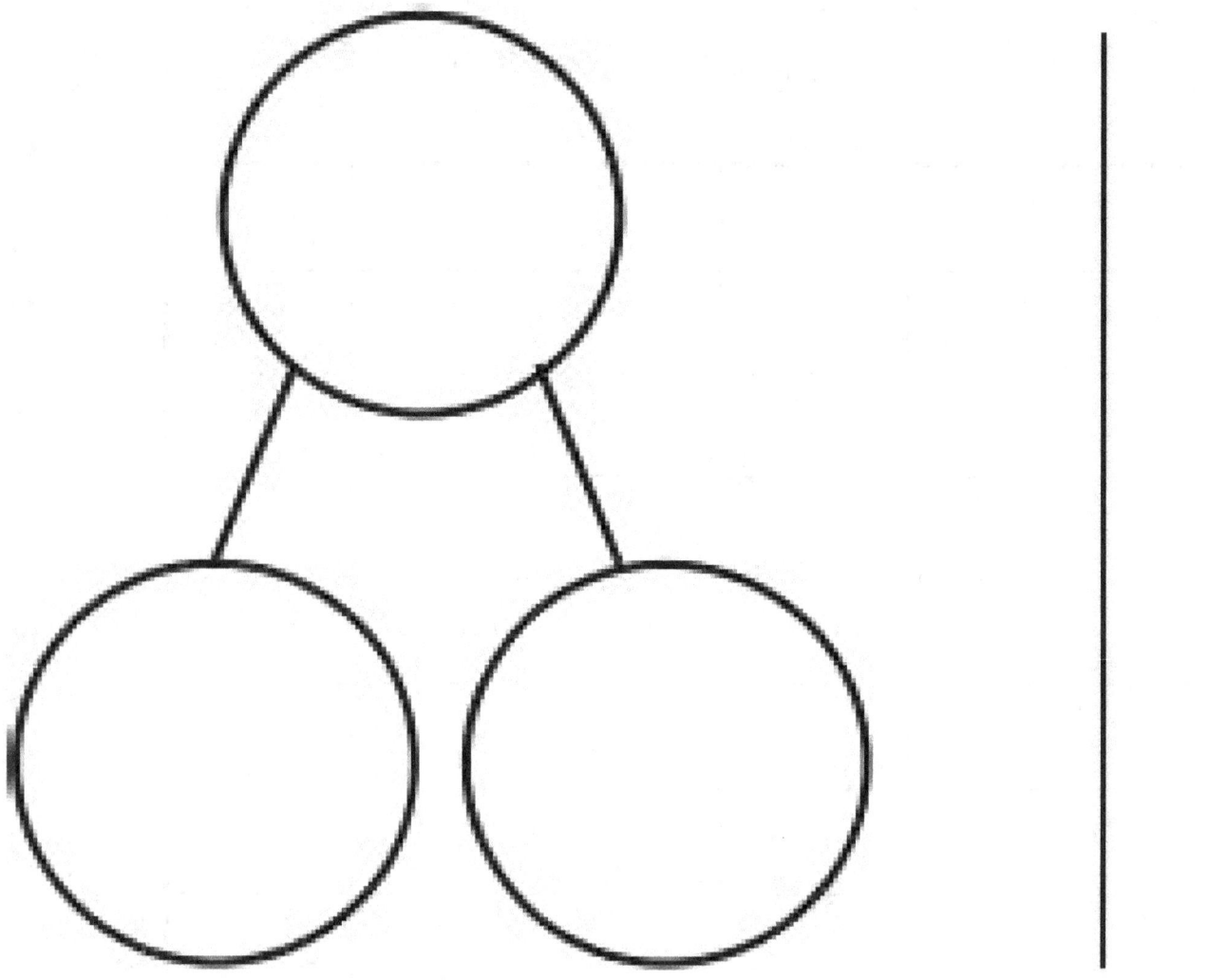

Break the total into parts. Write a number bond and addition and subtraction number sentences to match the story.

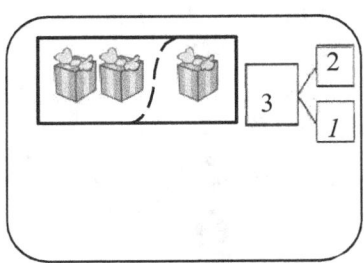

1. Rachel and Lucy are playing with 5 trucks. If Rachel is playing with 2 of them, how many is Lucy playing with?

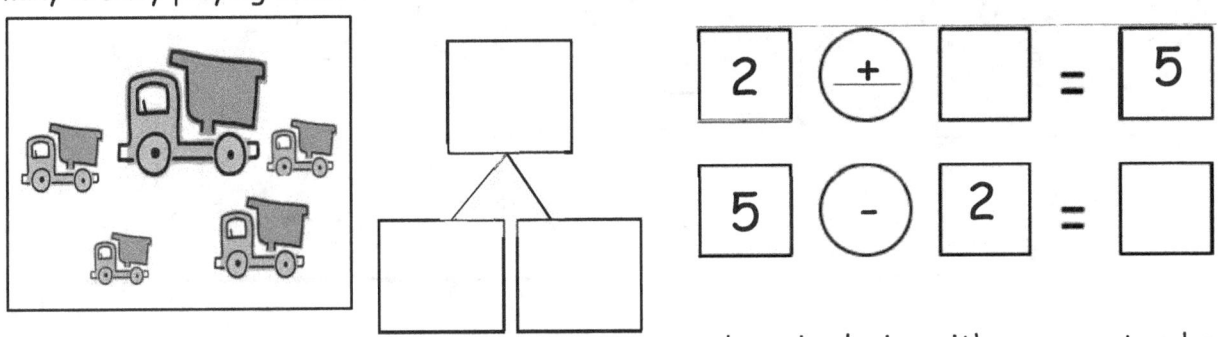

$2 \;+\; \square \;=\; 5$

$5 \;-\; 2 \;=\; \square$

Lucy is playing with _____ trucks.

---

2. Jane had 9 fish at the end of the day. She had 7 fish before she ate lunch. How many fish did she catch after lunch?

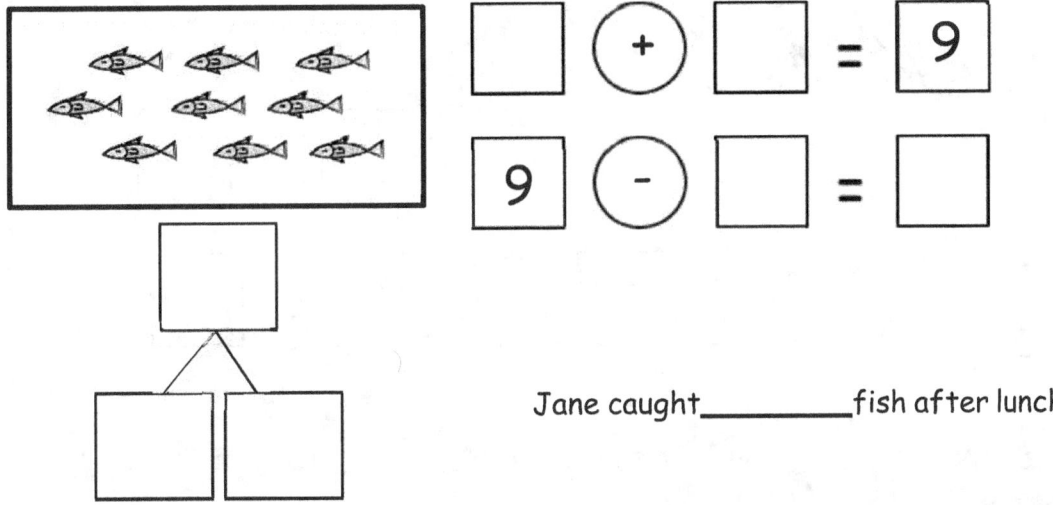

$\square \;+\; \square \;=\; 9$

$9 \;-\; \square \;=\; \square$

Jane caught _____ fish after lunch.

3. Dad bought 6 shirts. The next day he returned some of them. Now he has 2 shirts. How many shirts did Dad return?

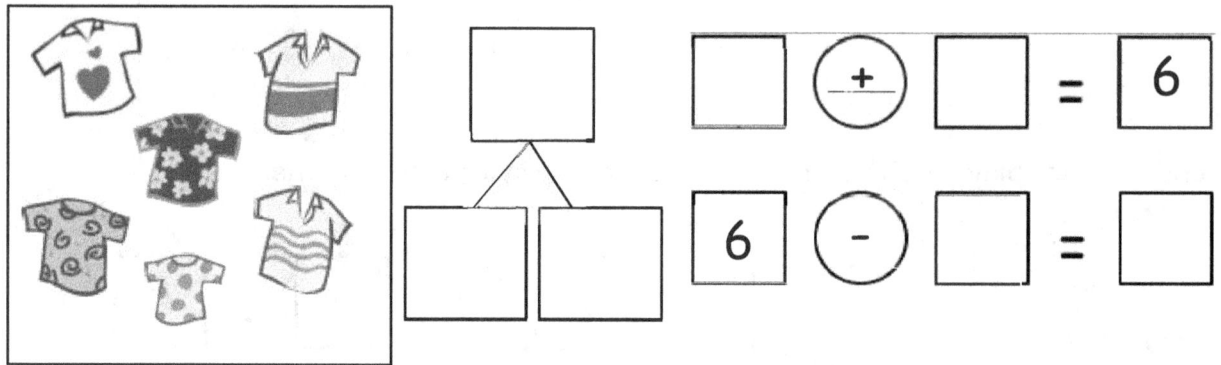

Dad returned_____shirts.

4. John had 3 strawberries. Then his friend gave him more fruit. Now John has 7 pieces of fruit. How many pieces of fruit did John's friend give him?

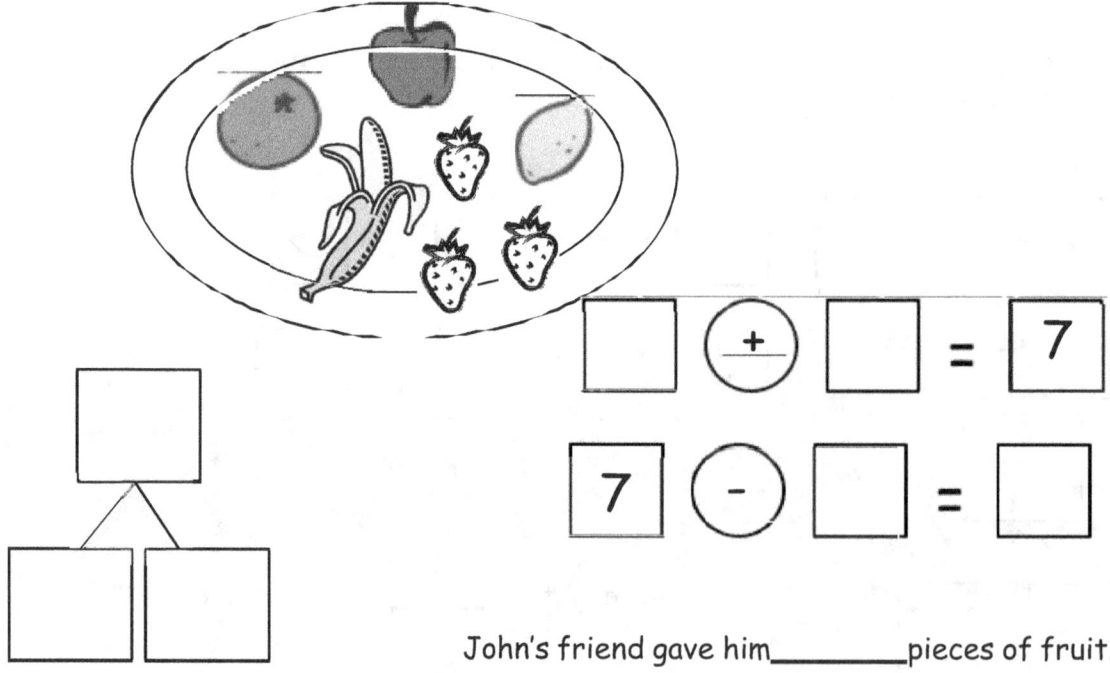

John's friend gave him_____pieces of fruit.

Solve the math story. Complete the number bonds and number sentences. Color the unknown number yellow.

Rich bought 6 cans of soda on Monday.

He bought some more on Tuesday.

Now he has 9 cans of soda.

How many cans did Rich buy on Tuesday?

Rich bought _____ cans.

□ + □ = □

□ - □ = □

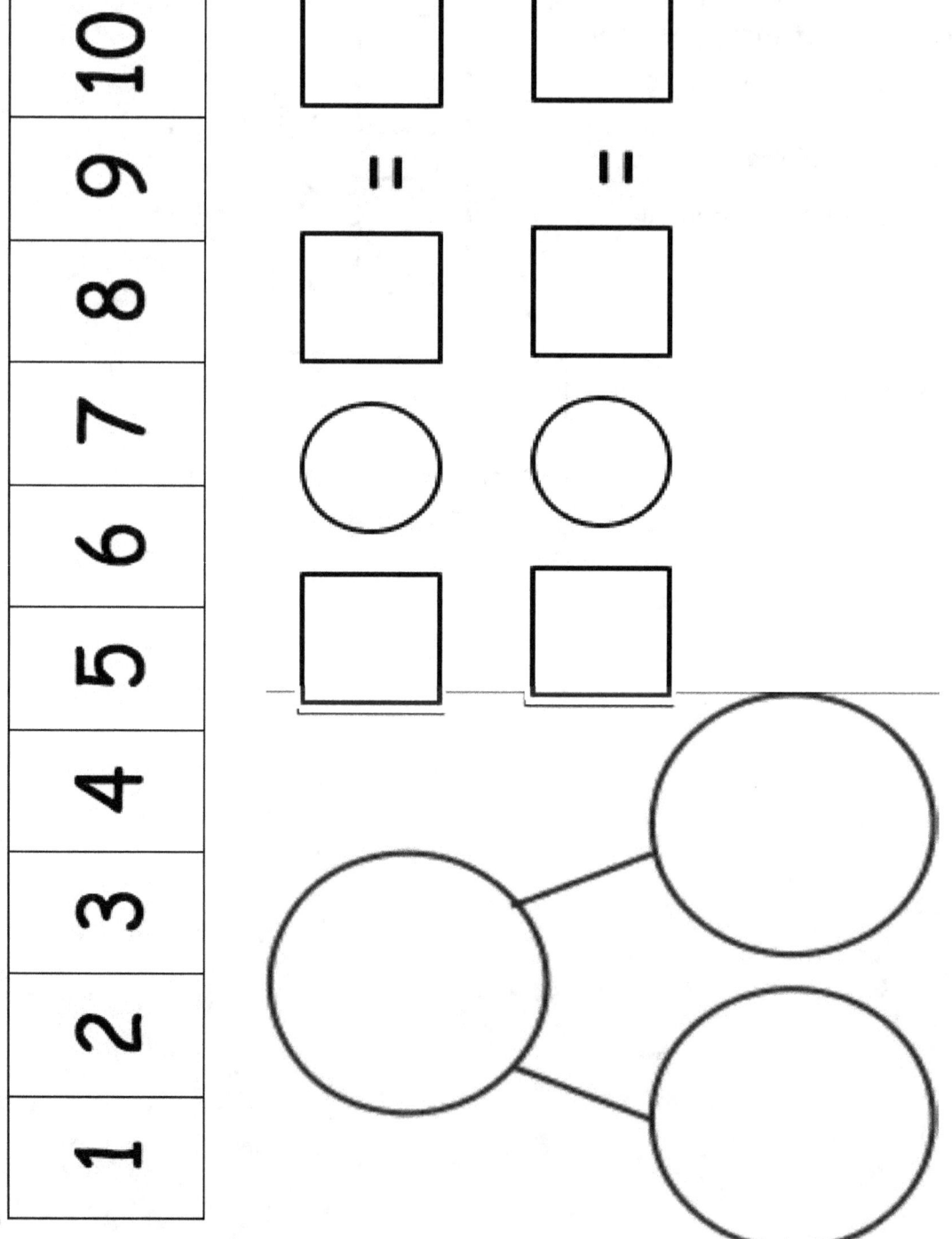

Use the number path to solve.

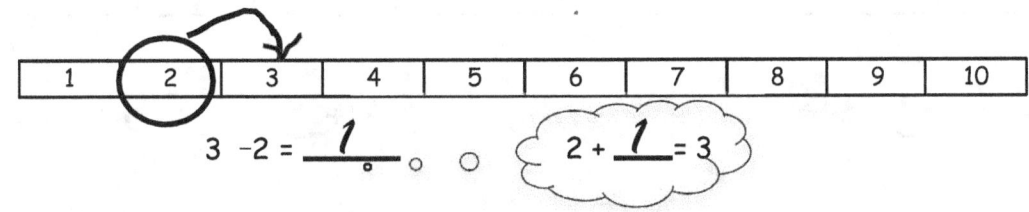

3 − 2 = __1__   2 + __1__ = 3

1. | 1 | 2 | 3 | 4 | 5 | 6 | 7 | 8 | 9 | 10 |

   6 − 4 = _____   4 + _____ = 6

2. | 1 | 2 | 3 | 4 | 5 | 6 | 7 | 8 | 9 | 10 |

   8 − 5 = _____   5 + _____ = 8

3. | 1 | 2 | 3 | 4 | 5 | 6 | 7 | 8 | 9 | 10 |

   9 − 6 = _____   6 + _____ =

4. | 1 | 2 | 3 | 4 | 5 | 6 | 7 | 8 | 9 | 10 |

   9 − 3 = _____   3 + _____ = 9

Use the number path to help you solve.

| 1 | 2 | 3 | 4 | 5 | 6 | 7 | 8 | 9 | 10 |

5. 5 − 4 = _____   4 + _____ = 5

6. 5 − 1 = _____   1 + _____ = 5

7. 7 − 5 = _____   5 + _____ = 7

8. 10 − 6 = _____   6 + _____ = 10

9. 9 − 3 = _____   3 + _____ = 9

Use the number path to solve. Write the addition sentence you used to help you solve.

| 1 | 2 | 3 | 4 | 5 | 6 | 7 | 8 | 9 | 10 |

a) 7 − 5 = _____   _____

b) 9 − 2 = _____   _____

c) _____ = 10 − 3   _____

Page Intentionally Left Blank

www.ingramcontent.com/pod-product-compliance
Lightning Source LLC
Chambersburg PA
CBHW060427220526
45465CB00008B/3040